The ESSE

DISCRETE
STRUCTURES

Mohammad Dadashzadeh, Ph.D.

Department of Decision Sciences
The Wichita State University, Wichita, Kansas

Research and Education Association
61 Ethel Road West
Piscataway, New Jersey 08854

THE ESSENTIALS® OF DISCRETE STRUCTURES

Printed in the United States of America

Library of Congress Catalog Card Number 91-62036

International Standard Book Number 0–87891–723-3

ESSENTIALS is a registered trademark of
Research and Education Association, Piscataway, New Jersey

WHAT "THE ESSENTIALS" WILL DO FOR YOU

This book is a review and study guide. It is comprehensive and it is concise.

It helps in preparing for exams, in doing homework, and remains a handy reference source at all times.

It condenses the vast amount of detail characteristic of the subject matter and summarizes the **essentials** of the field.

It will thus save hours of study and preparation time.

The book provides quick access to the important facts, principles, theorems, concepts, and equations in the field.

Materials needed for exams can be reviewed in summary form – eliminating the need to read and re-read many pages of textbook and class notes. The summaries will even tend to bring detail to mind that had been previously read or noted.

This "ESSENTIALS" book has been carefully prepared by educators and professionals and was subsequently reviewed by another group of editors to assure accuracy and maximum usefulness.

Dr. Max Fogiel
Program Director

CONTENTS

Chapter No.   **Page No.**

1   SET THEORY 1
1.1   Set Terminology 1
1.1.1   Types Of Sets 2
1.1.2   Cardinality 4
1.2   Set Operations 4
1.2.1   Set Laws 7
1.2.2   Cartesian Product 7

2   LOGIC 9
2.1   Propositional Calculus 9
2.2   Predicate Calculus.....................................14
2.3   Proof Techniques 15
2.3.1   Direct Proof..16
2.3.2   Contrapositive 16
2.3.3   Proof By Contradiction 17
2.3.4   Mathematical Induction 17

3   COMBINATORICS 20
3.1   Counting...20
3.2   Permutations 23
3.3   Combinations 24
3.4   Enumeration Theory 26
3.5   Partitions 27

**4   RELATIONS, MAPPINGS,
       AND FUNCTIONS** 29
4.1   Relations 29
4.1.1   Posets and Lattices.....................................30
4.2   Operations on Relations 34
4.3   Mappings and Functions 36

**5   ALGEBRAIC STRUCTURES AND
       HOMOMORPHISMS** 39
5.1   Operations 39
5.2   Groups, Rings and Fields, and Algebras......40
5.2.1   Boolean Algebra 41
5.3   Homomorphisms 44

6   GRAPHS AND TREES 46

6.1   Graph Terminology 46
6.2   Computer Representation of Graphs 52
6.3   Graph Algorithms 53

7   AUTOMATA THEORY 58

7.1   Finite State Machines 58
7.2   Push-Down and Linear Bounded Automata 61
7.3   Turing Machines 63

8   FORMAL LANGUAGES 66

8.1   Grammars and Languages 66
8.2   Classes of Grammars 68
8.3   Language Recognizers 70

9   COMPUTABILITY 73

9.1   Algorithms and Programs 73
9.2   The Universal Turing Machine 75
9.3   Unsolvability 75

10   SOME APPLICATION AREAS 79

10.1 Logic Circuit Design 79
10.2 Relational Databases and Relational Algebra 81

CHAPTER 1

SET THEORY

1.1   SET TERMINOLOGY

The simplest *discrete structure* is the set. A *set* is a collection of distinct objects of *any* kind. The distinct objects are called the *members* or *elements* of the set. There is no order to the set, nor is there any relationship among the members other than that they all belong to the set.

A set can be represented by listing its elements (separated by commas) between braces ({ and }). We may also give names to sets. A set is identified entirely by its distinct members, regardless of order or repetition, so that all the following are ways of identifying the same set.

{Joe's dog, "123," my house, 123, the Bible}

{Joe's dog, Joe's dog, 123, the Bible, my house, "123"}

{my house, 123, the Bible, 123, Joe's dog, 123, "123," my house}

In this set, the phrase "Joe's dog" should be taken to denote a *unique* object, namely a particular identifiable dog. Likewise, for the phrases "my house" and "the Bible." In the set, 123 denotes a unique abstract object, namely the number 123, and "123" denotes the unique character string "123." The important point is that members of a set must be all unique identifiable objects and that the set is completely determined by its members and not by the way in which it is defined – repetition and order are immaterial.

The fact that a set is completely determined when its members are specified is formalized by the *principle of extension* – two sets are equal

1

if and only if they have the same members.

Sets may contain many members. For example, the set $S = \{x : x$ is an odd natural number$\}$ will have a *countably infinite* number of members. Here, the set S is represented by specifying the property which characterizes the elements in the set.

The fact that we can describe a set in terms of a property is formalized by the *principle of abstraction* — given any set A and any property P, there exists a set B whose elements are precisely those members of A which have the property P.

It is convenient to name certain standard sets so that we can refer to them easily. Let,

N = the set of natural numbers: 1, 2, 3, ...

Z = the set of integers: ..., −2, −1, 0, 1, 2, ...

Q = the set of rational numbers

R = the set of real numbers

Since a set may contain as members objects of any kind, the members of a set may themselves be sets. For example, the set T = {123, {Joe's dog, "123," my house, 123, the Bible}} has two members one of which is a set. It is important to recognize that although "123" is an element of the set which is a member of T, "123" is *not* an element of T.

If a specific object a is a member of a set S, we write $a \; \varepsilon \; S$, where ε, the Greek letter epsilon, is the set membership operator. If a is not a member of S, we write $a \notin S$.

1.1.1 TYPES OF SETS

The set with no members is called the *empty set* or *null set* and is denoted by **0** or { }. Because a set is identified by its members, there is just one unique empty set: like all other sets, it is the same no matter how it is defined. ("The set of all odd numbers divisible by 2" and "the set of all spaceships with a maximum speed surpassing the speed of light" both define the empty set.)

If every member of a set S is also a member of a set T, then S is said to be a *subset* of T which is denoted by $S \subseteq T$. In addition, if there is at least one element of T that is not an element of S, then S is a *proper subset* of T, denoted by $S \subset T$. The following results are immediate consequences of this definition:

For any set A,

1. $0 \subseteq A$
2. $A \subseteq A$
3. if $A \subseteq B$ and $B \subseteq C$, then $A \subseteq C$
4. $A = B$ if and only if $A \subseteq B$ and $B \subseteq A$

The subset relationship can be illustrated by means of *Venn diagrams*. In a Venn diagram the *universal set* is represented by the interior of a rectangle, and the other sets are represented by regions (usually circles) within the rectangle. If $A \subseteq B$, then the region representing A will be entirely within the region representing B as in Figure 1a. If A and B are *disjoint*, then the regions representing them will be completely separated as in Figure 1b. Finally, if A and B have some common elements but neither is a subset of the other, this is illustrated as in Figure 1c.

Figures 1(a), (b), (c):

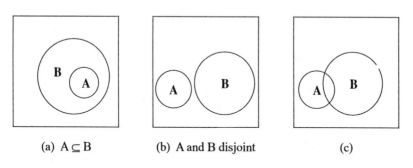

(a) $A \subseteq B$ (b) A and B disjoint   (c)

Given a set S we can form a new set whose elements are all of the subsets of S. This new set is called the *power set* of S and is denoted by $\wp(S)$. $\wp(S)$ will always include as members the empty set and the set S itself. For example, if $S = \{a, \{b, c\}, d\}$, then $\wp(S)$ consists of the following subsets of S: $0, \{a\}, \{\{b, c\}\}, \{d\}, \{a, \{b, c\}\}, \{a, d\}, \{\{b, c\}, d\}$, and $\{a, \{b, c\}, d\}$. Let S have n elements. Since each element of S may or may not belong to a subset of S, there will be exactly 2^n subsets of S, and thus $\wp(S)$ will have 2^n elements. (Note that $\wp(0)$ will have only one element, 0 itself.)

3

1.1.2 CARDINALITY

The *cardinality* of a finite set S, denoted $| S |$, is the number of its members. In a finite set we can always designate one element as the first member, another element as the second member, and so forth. If the set is infinite we may still be able to designate a first member, a second member, and so on through the set. Such infinite sets are said to be *denumerable* or *countably infinite* since we can count or enumerate all of their elements even though we cannot specify how many members they have. The set N, for example, is denumerable. So is the set Z since its members can be enumerated according to the following scheme: 0, -1, 1, -2, 2, -3, 3, etc.

We now show that the set of all the real numbers between 0 and 1 is *uncountable*. Consider that every member of this set may be written as a fraction of the form $0.d_1 d_2 d_3...$ where d_i is a digit between 0 and 9. Now let us assume that the set is countable and its members can be listed as follows:

$$0.d_{11} d_{12} d_{13} ...$$
$$0.d_{21} d_{22} d_{23} ...$$
$$0.d_{31} d_{32} d_{33} ...$$
$$...$$

We now construct a real number $p = 0.p_1 p_2 p_3...$ between 0 and 1 by letting $p_i = (d_{ii} + 1) \bmod 10$. In this manner, p will differ from the first number in the list at its first decimal digit, from the second number at its second decimal digit, and so on. Therefore, p cannot be the same as any number in the list. But the list was supposed to include all real numbers between 0 and 1. So, there is a contradiction and no such listing can exist. Hence, the set of all the real numbers between 0 and 1 is uncountable. The preceding proof is a very famous proof by contradiction known as *Cantor's diagonalization method*.

1.2 SET OPERATIONS

Just as arithmetic operations such as addition and multiplication allow us to construct new numbers from a given set of numbers, so we have a number of operations on sets which produce new sets. The *intersection* of two sets A and B, denoted $A \cap B$, consists of those elements

4

belonging to both A and B. The *union* of two sets A and B, denoted $A \cup B$, consists of those elements belonging to A, or B, or both. The *absolute complement* (or simply *complement*) of a set A, denoted A', consists of those elements belonging to the universal set $\mathbf{U}$ (or the universe of discourse) which are not members of A. The *relative complement* (or *difference*) of sets A and B, denoted $A - B$, consists of those elements of A which do not belong to B. These operations are pictorially represented by Venn diagrams as in Figure 2.

Figure 2:

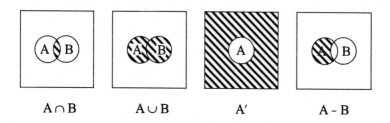

| $A \cap B$ | $A \cup B$ | A' | $A - B$ |

Venn diagrams can be used to illustrate the various possible operations between two, three, and four sets (Figures 3a, 3b, 3c).

Figure 3a:

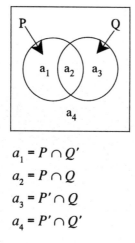

$$a_1 = P \cap Q'$$
$$a_2 = P \cap Q$$
$$a_3 = P' \cap Q$$
$$a_4 = P' \cap Q'$$

Figure 3b:

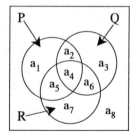

$$a_1 = P \cap Q' \cap R' \qquad\qquad a_2 = P \cap Q \cap R'$$
$$a_3 = P' \cap Q \cap R' \qquad\qquad a_4 = P \cap Q \cap R$$
$$a_5 = P \cap Q' \cap R \qquad\qquad a_6 = P' \cap Q \cap R$$
$$a_7 = P' \cap Q' \cap R \qquad\qquad a_8 = P' \cap Q' \cap R'$$

Figure 3c:

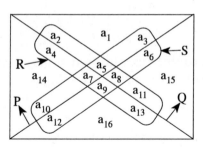

$$a_1 = P \cap Q \cap R' \cap S' \qquad\qquad a_9 = P' \cap Q' \cap R \cap S$$
$$a_2 = P \cap Q \cap R \cap S' \qquad\qquad a_{10} = P \cap Q' \cap R' \cap S$$
$$a_3 = P \cap Q \cap R' \cap S \qquad\qquad a_{11} = P' \cap Q \cap R \cap S'$$
$$a_4 = P \cap Q' \cap R \cap S' \qquad\qquad a_{12} = P' \cap Q' \cap R' \cap S$$
$$a_5 = P \cap Q \cap R \cap S \qquad\qquad a_{13} = P' \cap Q' \cap R \cap S'$$
$$a_6 = P' \cap Q \cap R' \cap S \qquad\qquad a_{14} = P \cap Q' \cap R' \cap S'$$
$$a_7 = P \cap Q' \cap R \cap S \qquad\qquad a_{15} = P' \cap Q \cap R' \cap S'$$
$$a_8 = P' \cap Q \cap R \cap S \qquad\qquad a_{16} = P' \cap Q' \cap R' \cap S'$$

1.2.1   SET LAWS

Under the above operations, sets satisfy *a* number of laws (or *set identities*) which are listed below. It is a fact of set theory, known as *principle of duality*, that the dual of each set identity is also an identity. The *dual identity* for a set identity is obtained by replacing each occurrence of $\cap$, $\cup$, U, and 0 in it by $\cup$, $\cap$, 0, and U respectively.

1. Commutative Laws: $P \cap Q = Q \cap P$ and its dual identity
$$P \cup Q = Q \cup P.$$

2. Associative Laws: $(P \cap Q) \cap R = P \cap (Q \cap R)$ and its dual identity $(P \cup Q) \cup R = P \cup (Q \cup R)$.

3. Distributive Laws: $P \cap (Q \cup R) = (P \cap Q) \cup (P \cap R)$ and its dual identity $P \cup (Q \cap R) = (P \cup Q) \cap (P \cup R)$.

4. De Morgan's Laws: $(P \cap Q)' = P' \cup Q'$ and its dual identity $(P \cup Q)' = P' \cap Q'$.

5. Idempotent Laws: $P \cap P = P$ and its dual identity $P \cup P = P$.

6. Complement Laws: $(P')' = P$; $P \cap P' = 0$ and its dual identity $P \cup P' = U$.

7. Identity Laws: $P \cap U = P$ and its dual identity $P \cup 0 = P$; $P \cup U = U$ and its dual identity $P \cap 0 = 0$.

1.2.2   CARTESIAN PRODUCT

The final set operation we introduce is perhaps the most important and powerful of the set operations. The *Cartesian product* (or *cross product*) of two sets A and B, denoted $A \times B$, is the set of all *ordered pairs* whose first component comes from A and whose second component comes from B. The ordered pairs are written as $<x,y>$ where x belongs to A and y belongs to B. For example, if $A = \{1, 2, 3\}$ and $B = \{a, b\}$, then the set $A \times B$ is equal to $\{<1,a>, <1,b>, <2,a>, <2,b>, <3,a>, <3,b>\}$ and the set $B \times A$ is the same as $\{<a,1>, <a,2>, <a,3>, <b,1>, <b,2>, <b,3>\}$. If $|A|$ is the cardinality of A and $|B|$ is the cardinality of B, then it is easy to show that both $A \times B$ and $B \times A$ contain $|A| \times |B|$ elements.

Similarly, the Cartesian product of the sets $A_1, A_2, ..., A_n$ is the set of all *n-tuples* $<a_1, a_2, ..., a_n>$ such that $a_i \, \varepsilon \, A_i$ for $i=1, 2, ..., n$. For example, given the sets $A_1 = \{a, b\}$, $A_2 = \{1, 2, 3\}$, and $A_3 = \{x, y\}$, the Cartesian product $A_1 \times A_2 \times A_3$ is the set $\{$<a,1,x>, <a,1,y>, <a,2,x>, <a,2,y>, <a,3,x>, <a,3,y>, <b,1,x>, <b,1,y>, <b,2,x>, <b,2,y>, <b,3,x>, <b,3,y>$\}$.

CHAPTER 2

LOGIC

2.1 PROPOSITIONAL CALCULUS

Propositional calculus is concerned with *propositions*, or *statements* that can be declared either to be true or to be false. The basic axioms about propositions are: 1) A proposition is either true or false, and 2) A proposition cannot be simultaneously true and false.

Let the letter p stand for the statement "I am taking Logic 101," and let q stand for "I am taking English 102." The symbols T and F will be assigned to a true and false statement, respectively. The basic logical operations (*connectives*) to form *compound statements* are: *conjunction* ($\wedge$), *disjunction* ($\vee$), *exclusive disjunction* ($\underline{\vee}$), and *negation* (~). Using the statements p and q we have:

Compound Statement	Stands for
$p \wedge q$	I am taking both Logic 101 and English 102.
$p \vee q$	I am taking at least one of the two courses (i.e., Logic 101, English 102, or both).
$p \underline{\vee} q$	I am only taking one of the courses, Logic 101 or English 102.
~p	I am not taking Logic 101.
~q	I am not taking English 102.
~$(p \wedge q)$	I am not taking both Logic 101 and English 102.

9

The last statement, ~(p ∧ q), implies three possibilities: p ∧ ~q, ~p ∧ q, and ~p ∧ ~q, which can be summarized as ~p ∨ ~q. Therefore, ~(p ∧ q) is equivalent to ~p ∨ ~q which can be formally demonstrated by constructing their *truth tables*.

p	q	$p \wedge q$	$\sim(p \wedge q)$	$\sim p$	$\sim q$	$\sim p \vee \sim q$
T	T	T	F	F	F	F
T	F	F	T	F	T	T
F	T	F	T	T	F	T
F	F	F	T	T	T	T

Since each of the two propositions p and q may take the values T and F independently of the other, there are four possible combinations of truth values: *TT, TF, FT,* and *FF*. For each of these possibilities a compound statement about p and q may take on either the value T or the value F. Therefore, there are $2^4 = 16$ possible *truth functions* to be defined for two propositions. The following table shows the possibilities.

p	q	1 logically true	2 $p \vee q$	3 $q \rightarrow p$	4 p	5 $p \rightarrow q$	6 q	7 $p=q$	8 $p \wedge q$	9 $\sim(p \wedge q)$	10 $\sim(p=q)$	11 $\sim q$	12 $\sim(p \rightarrow q)$	13 $\sim p$	14 $\sim(q \rightarrow p)$	15 $\sim(p \vee q)$	16 logically false
T	T	T	T	T	T	T	T	T	T	F	F	F	F	F	F	F	F
T	F	T	T	T	T	F	F	F	F	T	T	T	T	F	F	F	F
F	T	T	T	F	F	T	T	F	F	T	T	F	F	T	T	F	F
F	F	T	F	T	F	T	F	T	F	T	F	T	F	T	F	T	F

Let us interpret the significance of each column.

Column 1: represents a statement that is logically true regardless of the truth values of p and q. For example, if "I am taking Logic 101" (p), and "I am taking English 102" (q), then "2 + 2 = 4."

Column 2: represents the truth table for $p \vee q$.

Column 3: represents the truth table for $q \rightarrow p$ which can be stated in words as "q implies p" or "if q, then p" or "a sufficient condition for p to be true is for q to be true." Note that the condition is *not necessary* because p can be true when q is false. The operation $\rightarrow$ is named the *conditional* or *implication*, q is called the *antecedent* and p the *consequent*.

Column 4:	represents a restatement of p.

Column 4: represents a restatement of p.

Column 5: represents $p \to q$.

Column 6: represents a restatement of q.

Column 7: represents $p = q$ which can be stated in words as "if, and only if, p is true, then q is true" or "a necessary and sufficient condition for q is p" or "p and q are equivalent." Note that the position of p and q can be interchanged in all of the previous wordings. The operation $=$ is called *a biconditional* or *equivalence*.

Column 8: represents $p \wedge q$.

Column 9: is $\sim(p \wedge q)$ which is negation of column 8.

Column 10: is $\sim(p = q)$ which is negation of column 7 and also represents the truth table for $p \veebar q$.

Column 11: is $\sim q$ which is negation of column 6.

Column 12: is $\sim(p \to q)$; negation of column 5.

Column 13: is $\sim p$; negation of column 4.

Column 14: is $\sim(q \to p)$; negation of column 3.

Column 15: is $\sim(p \vee q)$; negation of column 2.

Column 16: is negation of column 1 and represents a statement that is logically false regardless of truth values for p and q. For example, if "I am taking Logic 101" (p), and "I am taking English 102" (q), then "2 + 2 = 5."

A *tautology* is a compound statement that is true regardless of the truth values of its constituent simple statements. For example, the compound statement $(p = q) = [(p \to q) \wedge (q \to p)]$ is a tautology as demonstrated by the truth table below:

		1	2	3	4	5
p	q	$p = q$	$p \to q$	$q \to p$	$(p \to q \wedge q \to p)$	$(p=q)=[(p \to q)\wedge(q \to p)]$
T	T	T	T	T	T	T
T	F	F	F	T	F	T
F	T	F	T	F	F	T
F	F	T	T	T	T	T

A *contradiction* is a compound statement that is false regardless of the truth values of its constituent simple statements. For example: $p \wedge \sim p$.

Two compound statements are said to be *logically equivalent* or equal if and only if they have identical truth tables. Using truth tables, we can show that the equivalence statements in the following laws hold true.

1. Commutative Laws: $(p \wedge q) = (q \wedge p)$ and $(p \vee q) = (q \vee p)$.

2. Associative Laws: $(p \wedge q) \wedge r = p \wedge (q \wedge r)$
   and $(p \vee q) \vee r = p \vee (q \vee r)$.

3. Distributive Laws: $p \wedge (q \vee r) = (p \wedge q) \vee (p \wedge r)$
   and $p \vee (q \wedge r) = (p \vee q) \wedge (p \vee r)$.

4. De Morgan's Laws: $\sim(p \wedge q) = (\sim p \vee \sim q)$
   and $\sim(p \vee q) = (\sim p \wedge \sim q)$.

5. Idempotent Laws: $(p \wedge p) = p$ and $(p \vee p) = p$.

6. Complement Laws: $(p \wedge \sim p) = F$ and $(p \vee \sim p) = T$
   and $\sim\sim p = p$.

7. Identity Laws: $(p \wedge T) = p$ and $(p \wedge F) = F$
   and $(p \vee T) = T$ and $(p \vee F) = p$.

A *syllogism* is a valid *argument* with two or more (*premises*) propositions and a *conclusion*. For example:

Major premise: All men are mortal. p

Minor premise: Socrates is a man. q

Conclusion: Therefore, Socrates is mortal. r

A basic function of logic is to check the validity of an argument, that is, the assertion that the conclusion follows from the premises (two or more). To do this, one must show that the compound statement representing conjunction of the premises implying the conclusion is a tautology. Consider the following example:

Premise 1: $p \to q$

Premise 2: p

Conclusion: q

To prove that the above argument is valid, it is sufficient to demonstrate that $[(p \rightarrow q) \wedge p] \rightarrow q$ is a tautology. This is accomplished by constructing the truth table below:

p	q	$p \rightarrow q$	$[(p \rightarrow q) \wedge p]$	$[(p \rightarrow q) \wedge p] \rightarrow q$
T	T	T	T	T
T	F	F	F	T
F	T	T	F	T
F	F	T	F	T

The truth table provides an effective test to determine the validity of arguments in the propositional calculus. However, it is also possible to *construct a proof* using the following common *axiom system* for the propositional calculus:

Axiom 1: $p \rightarrow (q \rightarrow p)$

Axiom 2: $(p \rightarrow (q \rightarrow r)) \rightarrow ((p \rightarrow q) \rightarrow (p \rightarrow r))$

Axiom 3: $(\sim p \rightarrow \sim q) \rightarrow ((\sim p \rightarrow q) \rightarrow p)$

Inference Rule 1   (modus ponens). From A and $A \rightarrow B$, infer B.

Inference Rule 2   (substitution). From string A, well-formed formula B, and propositional variable p, generate string C by substituting B for each occurrence of p in A.

By way of demonstration we show below the proof of the theorem: $p \rightarrow p$.

1. $(p \rightarrow (q \rightarrow r)) \rightarrow ((p \rightarrow q) \rightarrow (p \rightarrow r))$ ....Axiom 2

2. $(p \rightarrow ((p \rightarrow p) \rightarrow r)) \rightarrow ((p \rightarrow (p \rightarrow p)) \rightarrow (p \rightarrow r))$ ... Substitution of $(p \rightarrow p)$ for q

3. $(p \rightarrow ((p \rightarrow p) \rightarrow p)) \rightarrow ((p \rightarrow (p \rightarrow p)) \rightarrow (p \rightarrow p))$ ... Substitution of p for r

4. $p \rightarrow (q \rightarrow p)$ ....Axiom 1

5. $p \rightarrow ((p \rightarrow p) \rightarrow p)$ ....Substitution of $(p \rightarrow p)$ for q

6. $(p \rightarrow (p \rightarrow p)) \rightarrow (p \rightarrow p)$ ....Modus Ponens 3, 5

7. $p \rightarrow (p \rightarrow p)$ ....Substitution p for q in 4

8. $p \rightarrow p$ ....Modus Ponens 6, 7

The statements generated by applying the rules of inference from the axioms are, like the axioms themselves, tautologies, and are called *theorems*. Moreover, the *completeness theorem* states that a statement in the propositional calculus is a theorem (i.e., can be derived from the axioms by the rules of inference) if and only if it is a tautology.

2.2 PREDICATE CALCULUS

A propositional calculus statement cannot have a variable in it because the truth value of the statement becomes unspecified. For example, the expression "$x > 0$" does not have a fixed truth value even if we assume that x represents numeric values. However, expressions containing variables can be made into statements by adding *quantifiers*.

The *universal quantifier* is symbolized by an upside down A, $\forall$, and is read "for all," "for every," or "for each." Therefore, $(\forall x)(x > 0)$, is read "for all x, x is greater than zero." The truth value of this quantified expression *depends* on the collection of objects from which x may be chosen (called the *domain* of x). If the domain of x is integers then the statement is false. If the domain of x is odd natural numbers then the statement is true.

The predicate calculus (more formally, the *first-order predicate calculus*) is an extension of the propositional calculus which allows the use of variables denoting individual objects (but not sets of objects) and includes the *existential quantifier* (symbolized by a backwards E, $\exists$, and read "there exists one" or "for at least one") in addition to the universal quantifier. In expressions such as $(\forall x)(P(x))$ or $(\exists x)(Q(x))$, the symbols P and Q are called *predicates*, and x is called a *dummy argument* since the truth values of the expressions remain the same in a given interpretation if they are written, say, as $(\forall y)(P(y))$ or $(\exists z)(Q(z))$, respectively.

An *interpretation* for an expression involving quantifiers and predicates consists of the following: 1) a collection of objects called the *domain* of the interpretation which must include at least one object; 2) an assignment of a property of the objects in the domain to each predicate in the expression; and 3) an assignment of a par-ticular object in the domain to each constant symbol in the expres-sion. For example, the statement $(\forall x)(\forall y)(P(x,y) \rightarrow \sim P(y,x))$, under the interpretation that the domain of variables x and

14

y is the set of PEOPLE and that $P(x,y)$ is the binary predicate representing the PARENTHOOD relationship amongst PEOPLE (i.e., $P(x,y)$ means that x is parent of y), evaluates to true.

In predicate calculus the notion of tautology splits into the idea of a *valid* statement, which is true in all possible interpretations, and a *satisfiable* statement, which is true in at least one interpretation. Theorems in the predicate calculus correspond exactly to the valid statements, but there are statements that *cannot* be proven to be either a theorem or not to be a theorem. The basic problem lies in that, in the predicate calculus, testing whether a given statement is a theorem requires examining all possible domains, including the infinite ones, and there is no guarantee that such a test will ever come to an end. A number of such *undecidable* statements are known but their existence does not lessen the importance of quantification and the predicate calculus in computer science.

The following are useful theorems in the predicate calculus:

1. $(\forall x)(P(x)) = (\exists x)(\sim P(x))$
2. $\sim(\forall x)(\sim P(x)) = (\exists x)(P(x))$
3. $(\forall x)(\forall y)(P(x,y)) = (\forall y)(\forall x)(P(x,y))$
4. $(\exists x)(\exists y)(P(x,y)) = (\exists y)(\exists x)(P(x,y))$
5. $(\forall x)(P(x) \wedge Q(x)) = [(\forall x)(P(x))] \wedge [(\forall x)(Q(x))]$
6. $(\exists x)(P(x) \vee Q(x)) = [(\exists x)(P(x))] \vee [(\exists x)(Q(x))]$

Note, however, that it is *not* the case that: $(\forall y)(\exists x)(P(x,y)) = (\exists x)(\forall y)(P(x,y))$. For, consider the interpretation where the domain of variables x and y is the set of PEOPLE and that $P(x,y)$ is the binary predicate representing the PARENTHOOD relationship amongst PEOPLE, under which $(\forall y)(\exists x)(P(x,y))$ translates to "each person has a parent" while $(\exists x)(\forall y)(P(x,y))$ would translate to "there is a person who is the parent of all persons."

2.3   PROOF TECHNIQUES

Suppose you need to prove a conjecture $P \rightarrow Q$. Although a single *counter-example* is sufficient to refute (disprove) a conjecture, in general providing more and more examples for which the assertion is true does not prove the conjecture. The only exception is when your conjecture is an assertion about a finite set, in which case you can prove the conjecture by demonstrating that it is true for every member of the set.

2.3.1 DIRECT PROOF

The obvious approach to proving $P \rightarrow Q$ is the *direct proof* - assume the hypothesis P and deduce the conclusion Q. For example, consider the following proof of the assertion that the product of two odd integers is also an odd integer.

Hypotheses: x and y are integers; x is odd; y is odd.

1. $x = 2m + 1$ .... definition of an odd integer and hypothesis
2. $y = 2n + 1$ .... definition of an odd integer and hypothesis
3. $xy = (2m + 1)(2n + 1)$ .... property of equivalence
4. $xy = 2(2mn + m + n) + 1$ .... properties of addition and multiplication of integers
5. $xy = 2k + 1$ .... substitution of k for $(2mn + m + n)$
6. xy is odd .... definition of an odd integer

2.3.2 CONTRAPOSITIVE

It may sometimes be easier to prove the *contrapositive* of $P \rightarrow Q$, that is to prove $\sim Q \rightarrow \sim P$. This technique is called *proof by contraposition* and takes advantage of the tautology: $(\sim Q \rightarrow \sim P) \rightarrow (P \rightarrow Q)$. For example, consider the following proof of the assertion that if xy is odd then both x and y must be odd.

To prove: xy is odd $\rightarrow$ [(x is odd) $\wedge$ (y is odd)]

Proof by contraposition: We show that $\sim$[(x is odd) $\wedge$ (y is odd)] $\rightarrow \sim$(xy is odd).

1. By De Morgan's law we need to show: $\sim$(x is odd) $\vee \sim$(y is odd) $\rightarrow \sim$(xy is odd)

2. Or equivalently, (x is even) $\vee$ (y is even) $\rightarrow$ (xy is even) .... definition of odd and even

3a. if x is even and y is odd, then

$$x = 2m; y = 2n + 1$$
$$xy = (2m)(2n + 1) = 2(2mn + m); xy \text{ is even}$$

3b. if x is odd and y is even, then

$$x = 2m + 1; y = 2n$$
$$xy = (2m + 1)(2n) = 2(2mn + n); xy \text{ is even}$$

3c. if x is even and y is even, then

$$x = 2m; y = 2n$$
$$xy = (2m)(2n) = 2(2mn); xy \text{ is even}$$

2.3.3   PROOF BY CONTRADICTION

Another proof technique is *proof by contradiction* where you assume both the hypothesis and the negation of the conclusion and then try to deduce a contradiction. The technique takes advantage of the tautology: $(P \wedge \sim Q \rightarrow 0) \rightarrow (P \rightarrow Q)$. For example, consider the following proof of the assertion that *given a set A, there exists a set B such that B is not a member of A.*

Proof: Consider the property P: "X is a set and $X \notin X$". By the *principle of abstraction*, there exists a set B whose elements are exactly those elements of A which have the property P. That is, $B = \{ X \in A : X \text{ is a set and } X \notin X \}$. We now show that the set B does not belong to the set A. To do that consider that for the set B there are two possibilities: (i) $B \in B$; or (ii) $B \notin B$.

Suppose that $B \in A$. Under the possibility (i), we have $B \in A$ and $B \in B$, and therefore by definition of B, we have $B \notin B$, contradicting (i). Under the possibility (ii), we have $B \in A$ and $B \notin B$, and therefore by definition of B, we have $B \in B$, contradicting (ii). Hence, the assumption that $B \in A$ is necessarily false.

2.3.4   MATHEMATICAL INDUCTION

Finally, *mathematical induction* is a proof technique that is especially useful in computer science. It can be used to prove the truth of some property with respect to all members of a set whose members can be constructed by a finite number of steps from a finite number of initial members. This is more simply put if we keep to natural numbers as the set in question. The *induction principle* states that to prove a property P holds for all positive integers, we must show $P(1)$ (i.e., 1 has the property P) and that for any positive integer k, $P(k) \rightarrow P(k+1)$ (i.e., if any number has the property P, then so does the next number). The following example illustrates proof by mathematical induction:

Example: Prove that $1 + 3 + 5 + ... + (2n - 1) = n^2$ is true for any positive integer n.

Proof by mathematical induction:

Basis: $P(1)$ is true. $1 = 1^2$.

Induction Hypothesis: Assume $P(k)$ is true. That is, $1 + 3 + 5 + ... + (2k - 1) = k^2$.

Induction Step: Show $P(k + 1) = (k + 1)^2$.

$P(k + 1) =$

$1 + 3 + 5 + ... + (2k - 1) + (2(k+1) -1) =$

$k^2 + (2(k+1) - 1) =$

$k^2 + (2k + 1) =$

$(k + 1)^2$

It is important to remember that in a proof by mathematical induction, the proof of the induction step should be valid for *any* positive integer k. Failure to account for this may lead to surprising (and incorrect) results. Consider the following (invalid) theorem and its (incorrect) proof shown below:

Theorem: Let a be any positive number; for all positive integers n we have $a^{n-1} = 1$.

Proof by mathematical induction:

Basis: $P(1)$ is true. $a^{n-1} = a^{1-1} = a^0 = 1$.

Induction Hypothesis: Assume $P(i)$ is true for $i=1, 2, ..., k$. In particular we have: $a^{k-1} = 1$ and $a^{(k-1)-1} = a^{k-2} = 1$.

Induction Step: Show $P(k + 1)$ is also true.

$a^{(k+1)-1} = a^k = a^{k-1} / a^{-1} = (a^{k-1} \times a^{k-1}) / (a^{k-1} \times a^{-1}) = (a^{k-1} \times a^{k-1}) / (a^{k-2}) =$

$(1 \times 1) / 1 = 1$.

The problem with the preceding proof is that the proof of the induction step is not valid for $k = 1$, unless $a^{k-2} = a^{-1} = 1$, which would be true if and only if $a = 1$, in which case the theorem is indeed valid.

CHAPTER 3

COMBINATORICS

3.1   COUNTING

Combinatorics is that branch of mathematics that deals with counting the number of elements in a finite set. Many counting problems can be solved by applying the *multiplication principle* which states that if an event can occur in n ways and another in m ways, then there are $n \times m$ ways in which these two events can occur.

> **Example:**   A lotto ticket must consist of 6 unique numbers each of which must be between 1 and 44. How many lotto tickets are there?

We can think of the number of lotto tickets as the total number of possible outcomes of the sequence of events of choosing the first number, then the next number, etc. The first number can be any of the 44 numbers (1– 44) while the second number can be any of the 43 remaining numbers. Therefore, there are $44 \times 43 \times 42 \times 41 \times 40 \times 39 = 4096 \times 10^6$ different lotto tickets.

The *addition principle* which states that if A is the union of two disjoint sets B and C, and if B has n elements and C has m elements, then A has $n + m$ elements, is useful whenever we want to count the total number of outcomes for an event that can be broken into disjoint cases.

> **Example:**   An identifier in some dialects of BASIC must be either a single letter or a single letter followed by a single digit. How many such identifiers are possible?

The total number of identifiers is equal to the sum of the total number of identifiers consisting of a single letter (26) and the total number of identifiers consisting of a single letter followed by a single digit (26 × 10), or 286.

We conclude this section by examining the powerful *inclusion and exclusion principle*. Suppose we have a set of N objects with three properties a_1, a_2, and a_3. The properties are not mutually exclusive, that is, an object can have one or more of these properties. Let $N(a_i)$ denote the number of objects with property a_i, $N(a_i')$ the number of objects which do not have property a_i, $N(a_i a_j)$ the number of objects with both properties a_i and a_j, and $N(a_i'a_j')$ the number of objects with neither property a_i nor property a_j, and so forth. Now consider the Venn diagram in Figure 4a where the universe is the set of N objects and the circle marked a_1 is the set of objects with property a_1. Let the label placed in each region refer only to that small region. In other words, let us distinguish between the bounded region a_1 shaded in Figure 4b, and the set of all elements counted in $N(a_1)$ shaded in Figure 4c. Similarly, $N(a_2 a_3)$ counts both the bounded region $A_2 A_3$ and the bounded region $A_1 A_2 A_3$. Finally, the set of objects with none of the properties is the region in the rectangle outside the 3 circles denoted by O.

Let $|A_1|$ stand for the number of elements in the bounded region A_1. We have:

$$N = |O| + |A_1| + |A_2| + |A_3| + |A_1A_2| + |A_2A_3| + |A_1A_3| + |A_1A_2A_3|$$
$$N(a_1) = |A_1| + |A_1A_2| + |A_1A_3| + |A_1A_2A_3|$$
$$N(a_2) = |A_2| + |A_1A_2| + |A_2A_3| + |A_1A_2A_3|$$
$$N(a_3) = |A_3| + |A_1A_3| + |A_2A_3| + |A_1A_2A_3|$$
$$N(a_1a_2) = |A_1A_2| + |A_1A_2A_3|$$
$$N(a_1a_3) = |A_1A_3| + |A_1A_2A_3|$$
$$N(a_2a_3) = |A_2A_3| + |A_1A_2A_3|$$
$$N(a_1a_2a_3) = |A_1A_2A_3|$$

Now, if we perform the following calculation:

$$N - N(a_1) - N(a_2) - N(a_3) + N(a_1a_2) + N(a_2a_3) + N(a_1a_3) - N(a_1a_2a_3),$$

then we have $|O|$ counted once (in N); $|A_1|$ counted zero times (+1 in N, -1 in $N(a_1)$); $|A_2|$ counted zero times (+1 in N, -1 in $N(a_2)$); $|A_3|$ counted zero times (+1 in N, -1 in $N(a_3)$); $|A_1A_2|$ counted zero times (+1 in N,

-1 in $-N(a_1)$, -1 in $-N(a_2)$, $+1$ in $N(a_1a_2)$); $|A_1A_3|$ counted zero times ($+1$ in N, -1 in $-N(a_1)$, -1 in $-N(a_3)$, $+1$ in $N(a_1a_3)$); $|A_2A_3|$ counted zero times ($+1$ in N, -1 in $-N(a_2)$, -1 in $-N(a_3)$, $+1$ in $N(a_2a_3)$); $|A_1A_2A_3|$ counted zero times ($+1$ in N, -1 in $N(a_1)$, -1 in $N(a_2)$, -1 in $N(a_3)$, $+1$ in $N(a_1a_2)$, $+1$ in $N(a_1a_3)$, $+1$ in $N(a_2a_3)$, -1 in $N(a_1a_2a_3)$). Therefore, the total is just $|O|$ so that:

$$N(a_1'a_2'a_3') = N - N(a_1) - N(a_2) - N(a_3) + N(a_1a_2) + N(a_2a_3) + N(a_1a_3) - N(a_1a_2a_3).$$

Figure 4a:

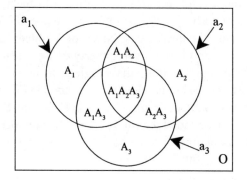

Figure 4b:

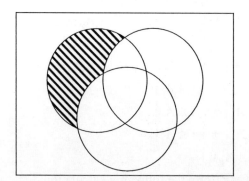

Figure 4c:

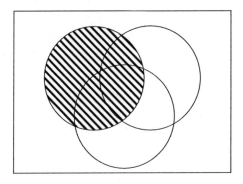

Example: How many numbers between 1 and 300 are not divisible by 2, 5, or 7?

Let a_1, a_2, and a_3 stand for the properties that a number between 1 and 300 is divisible by, respectively, 2, 5, and 7. We have: $N(a_1) = \lfloor 300/2 \rfloor = 150$, $N(a_2) = \lfloor 300/5 \rfloor = 60$, $N(a_3) = \lfloor 300/7 \rfloor = 42$, $N(a_1 a_2) = \lfloor 300/10 \rfloor = 30$, $N(a_2 a_3) = \lfloor 300/35 \rfloor = 8$, $N(a_3 a_1) = \lfloor 300/14 \rfloor = 21$, and $N(a_1 a_2 a_3) = \lfloor 300/70 \rfloor = 4$.

Then $N(a_1' a_2' a_3') = 300 - (150 + 60 + 42) + (30 + 8 + 21) - 4 = 103$.

Example: How many numbers between 1 and 300 are not divisible by 2 and 5 but are divisible by 7?

With a_1, a_2, and a_3 as in the previous example, from the Venn diagram in Figure 4 it is apparent that:

$N(a_1' a_2' a_3) = N(a_3) - N(a_3 a_1) - N(a_3 a_2) + N(a_3 a_2 a_1) = 42 - 8 - 21 + 4 = 17$.

3.2   PERMUTATIONS

A permutation is a particular *ordered* arrangement of the elements in a set. For example, given 3 objects, say a, b, and c, there are six possible permutations involving them: $abc, cba, acb, bca, bac, cab$. In general, the number of permutations of a set of cardinality n is $n!$ (read n *factorial*). For a positive integer n, $n!$ is defined as the product $n(n-1)(n-2)...1$, and $0!$ is defined to be 1.

An *ordered selection* of r distinct elements from a set of n elements

23

is called the *r-permutation of n objects* and is denoted by $P(n, r)$. To compute $P(n, r)$, we apply the multiplication principle and note that the first object chosen may be any one of the n objects, while the next object may be any of the $n - 1$ remaining. This pattern continues until r objects have been chosen. Thus, $P(n, r) = n \times (n-1) \times (n-2) \times \ldots \times (n - r + 1)$, or equivalently, $P(n, r) = n! / (n - r)!$.

Example: The number of ordered arrangements of 3 objects, a, b, and c, taken 2 at a time is given by: $3! / (3 - 2)! = 6$. They are: ab, ba, ac, ca, bc, cb.

Example: The number of ordered arrangements of 3 objects, a, b, and c, taken 2 at a time *with unlimited repetition* is by the multiplication principle $3 \times 3 = 9$. They are: aa, ab, ac, ba, bb, bc, ca, cb, cc.

The number of arrangements of n objects of which exactly q_1 are of one kind, q_2 are of a second kind, ..., and q_k are of the kth kind is given by: $n! / (q_1! \, q_2! \ldots q_k!)$.

Example: In how many different ways can 8 coins be chosen from 7 quarters and 5 dimes?

We have the following possibilities. The eight coins selected consist of 7 quarters and 1 dime, or 6 quarters and 2 dimes, or 5 quarters and 3 dimes, or 4 quarters and 4 dimes, or 3 quarters and 5 dimes. By the addition principle, the answer is:

$$(8! / (7!1!)) + (8! / (6!2!)) + (8! / (5!3!)) + (8! / (4!4!)) + (8! / (3!5!)) = 218.$$

3.3 COMBINATIONS

An *unordered* selection of r objects from a set of n objects (without repetition) is called a combination of n objects taken r at a time. The number of such combinations is denoted by $C(n, r)$ and is given by: $n! / (r!(n - r)!)$. Since any combinations of n objects taken r at a time determines $r!$ permutations of the objects in the combination, we can conclude that $P(n, r) = r! \times C(n, r)$.

Example: How many different poker hands (set of 5 cards) can be dealt from a 52-card deck?

Since the order of the cards does not matter, the answer is given by

C (52, 5) = 52! / (5!47!) = 2,598,960.

Example: How many poker hands can be formed with no two cards having the same face value?

We can choose the 5 distinct face values out of 13 possible ones in $C(13, 5)$ different ways. Then we can choose the suit for *each* card in 4 different ways. Therefore, the answer is: $4^5 \times C(13, 5) = 4^5 \times (13! / 8!5!) = 4^5 \times 13 \times 11 \times 9$.

Example: Four scientists are working on a project and wish to lock up project material in a safe at the end of the day. What is the minimum number of locks necessary if they require that, at a minimum, 3 of them be present before the safe can be opened? How many keys should each scientist carry?

For any combination of 2 scientists there should be at least one lock that they cannot open. Therefore, the minimum number of locks necessary is $C(4, 2) = 6$. The number of keys each scientist should carry can be determined by noting that a particular scientist must carry the key for each lock that cannot be opened by any combination of 2 other scientists. There are $C(3, 2)$ such combinations and thus the minimum number of keys to carry would be 3.

The number of combinations of n objects taken r at a time *with repetitions allowed* is $C(n + r - 1, r)$.

Example: Consider the 5 objects a, b, c, d, and e. The number of combinations of these 2 at a time with repetition is $C(5 + 2 - 1, 2) = C(6, 2) = 15$. The possibilities are: *aa, ab, ac, ad, ae, bb, bc, bd, be, cc, cd, ce, dd, de, ee.*

Example: Suppose we have n gifts and 3 children, $n \geq 3$. The children are considered distinct. If the gifts are considered distinct, then (*i*) in how many ways can exactly one gift be given to each child? (*ii*) in how many ways can all the gifts be given to the three children if each child must receive at least one gift?

Considering the first question. There are n choices for the first child, $n - 1$ for the second, and $n - 2$ for the third. Thus, the answer is $n(n - 1)(n - 2)$.

For the second question, consider that there are 3 choices for each gift and thus a total of 3^n ways of distributing the n distinct gifts amongst the three children. However, of these there will be some distributions in which one of the children does not receive any gift. The number of such distributions is obtained by $C(3, 2) \times 2^n$, where $C(3, 2)$ represents the number of ways of selecting the two children who would receive all the gifts, and 2^n represents the number of ways of distributing the n gifts amongst them. Finally, because the 3 possible distributions in which a particular child gets all of the n gifts is accounted for once by 3^n and twice by $C(3, 2) \times 2^n$, the final answer becomes: $3^n - C(3, 2) \times 2^n + 3$.

Example: Consider the situation in the previous example, but assume that the gifts are identical.

Now, since the gifts are identical, there is only one way to assign exactly one gift to each child. For the second question, consider that we can first give each child a gift and then assign the remaining $n-3$ gifts by choosing a child to receive the gift. Since repetition is allowed, i.e., the same child may be chosen more than once, the number of ways to assign the remaining $n-3$ gifts between the three children is given by the number of combinations of 3 objects taken $n-3$ at a time: $C(3 + (n-3) -1, n-3) = C(n-1, n-3)$.

We conclude this section by noting that the numbers $C(n, r)$ are called *binomial coefficients* because they appear as the coefficients in the expansion of $(a + b)^n$:

$$(a + b)^n = a^n + C(n, 1) \times a^{n-1} \times b + \ldots + C(n, r) \times a^{n-r} \times b^r + \ldots + b^n.$$

Binomial coefficients can also be computed using the following *recurrence relation* known as Pascal's formula, $C(n, r) = C(n-1, r) + C(n-1, r-1)$, with initial conditions that $C(n, 0) = 1$ and $C(n, n) = 1$.

3.4   ENUMERATION THEORY

The general procedure for counting permutations or combinations is based on the concept of a *generating function* - a function whose coefficients are really the values we seek. For combinations, the basic observation is that the values $C(n, r)$ occur naturally as coefficients in the binomial expansion $(1 + x)^n = \Sigma_{r=0,n} C(n, r) \times x^r$. The interpretation of the left side of this formula is that we are dealing with n objects (the exponent), each of which may be selected once (the $x = x^1$ term) or not at all (the $1 = x^0$ term). The interpretation of the right hand side of the formula is that the coefficient of x^r term is the number of ways r objects may be chosen from the n objects. This interpretation generalizes easily as the following example illustrates.

Example: In how many ways can we select 8 coins from a set of 2 pennies, 3 nickels, 4 dimes, and 5 quarters subject to the constraint that the number of quarters selected be odd?

The generating function for the number of pennies is $(1 + x + x^2)$, representing the fact that we can select either no pennies, or 1, or 2. For the nickels, the generating function is $(1 + x + x^2 + x^3)$, and for the dimes $(1 + x + x^2 + x^3 + x^4)$. The generating function for the number of quarters is $(x + x^3 + x^5)$. Therefore, the coefficient of x^8 in the expansion of $(1 + x + x^2)(1 + x + x^2 + x^3)(1 + x + x^2 + x^3 + x^4)(x + x^3 + x^5)$ provides the answer.

3.5   PARTITIONS

The number of partitions of a positive integer n, denoted by $p(n)$, is the number of ways of writing n as a sum of positive integers. For example, the seven partitions of 5 are: $5, 4 + 1, 3 + 2, 3 + 1 + 1, 2 + 2 + 1, 2 + 1 + 1 + 1$, and $1 + 1 + 1 + 1 + 1$. In such a partition as $3 + 2$, the numbers 3 and 2 are called the summands. In counting the number of partitions, the order of summands is irrelevant.

Let $p_k(n)$ denote the number of partitions of n with summands no larger than k. The following relations can be established:

$p(n) = p_n(n) = p_{n+1}(n) = p_{n+2}(n) = ...$,

$p_k(n) = p_{k-1}(n) + p_k(n - k)$ for $1 < k < n$,

$p_1(n) = 1$, and

$p_n(n) = 1 + p_{n-1}(n)$.

With these results it is a simple matter to make a table of values for $p_k(n)$.

$p_k(n)$	$k = 1$	$k = 2$	$k = 3$	$k = 4$	$k = 5$	$k = 6$
$n = 1$	1	1	1	1	1	1
$n = 2$	1	2	2	2	2	2
$n = 3$	1	2	3	3	3	3
$n = 4$	1	3	4	5	5	5
$n = 5$	1	3	5	6	7	7
$n = 6$	1	4	7	9	10	11

Let a, b, c, and d be unequal positive integers. Then the coefficient of x^n in the expansion of $(1 + x^a + x^{2a} + x^{3a} + ...)(1 + x^b + x^{2b} + x^{3b} + ...)(1 + x^c + x^{2c} + x^{3c} + ...)(1 + x^d + x^{2d} + x^{3d} + ...)$ equals the number of partitions of n with summands restricted to $a, b, c,$ and d. Each factor in the expansion must include all exponents not exceeding n.

Example: In how many ways is it possible to break a dollar bill into change?

Since coins come in the denominations 1, 5, 10, 25, and 50 cents, our task is to find the number of partitions of 100 with summands restricted to 1, 5, 10, 25, and 50. Therefore, the answer to the question may be obtained by determining the coefficient of x^{100} in the following expansion: $(1 + x + x^2 + ... + x^{100}) \times (1 + x^5 + x^{10} + ... + x^{100}) \times (1 + x^{10} + x^{20} + ... + x^{100}) \times (1 + x^{25} + x^{50} + x^{75} + x^{100}) \times (1 + x^{50} + x^{100})$. The answer is 292, and so there are 292 different ways of changing a dollar bill.

CHAPTER 4

RELATIONS, MAPPINGS, AND FUNCTIONS

4.1 RELATIONS

Let A and B be sets. A *binary* relation R from A to B is a subset of the Cartesian product $A \times B$. If $(a,b) \, \varepsilon \, R$, we say that a is related to b by R and denote this by $a \, R \, b$. The *domain* of R, a subset of A, is the set of all first elements in the ordered pairs that make up R. Similarly, the *range* of R, a subset of B, is defined to be the set of all second elements in the ordered pairs that make up R. If R is a relation from A to A, i.e., $R \subseteq A \times A$, then we say that R is a relation *on A*.

> **Example:** Let A be the set of all cities in the U.S., and B be the set of all the states in the U.S. Then, a relation "is a city in" can be defined from A to B such that if $(a, b) \, \varepsilon \, R$, then a must be a city in the state b.

For binary relations on a set A there are a number of useful and interesting properties. A relation R is *reflexive* if $\forall a \, \varepsilon \, A$, $a \, R \, a$. A relation R is *symmetric* if, whenever we have $a_1 \, R \, a_2$ then we also have $a_2 \, R \, a_1$. More formally, R is symmetric if $\forall a_1, a_2 \, \varepsilon \, A$, $a_1 \, R \, a_2 \to a_2 \, R \, a_1$. A relation R is *transitive* if, whenever we have $a_1 \, R \, a_2$ and $a_2 \, R \, a_3$ then we also have $a_1 \, R \, a_3$. Formally, R is transitive if $\forall a_1, a_2, a_3 \, \varepsilon \, A$, $a_1 \, R \, a_2 \, {}^\wedge \, a_2 \, R \, a_3 \to a_1 \, R \, a_3$.

> **Example:** On a population of people, the relation "is a sibling of" is transitive and symmetric but not reflexive since one

is not normally regarded as one's own sibling. The relation "is a sister of" is transitive but not reflexive nor symmetric. The relation "is a parent of" is neither reflexive, nor symmetric, nor transitive.

Example: The relation $\subseteq$ on sets is reflexive because for all sets S, $S \subseteq S$. It is also transitive because for all sets S_1, S_2, S_3, if $S_1 \subseteq S_2$ and $S_2 \subseteq S_3$ then $S_1 \subseteq S_3$. However, $\subseteq$ is not symmetric because if $S_1 \subseteq S_2$ then it is not necessarily the case that $S_2 \subseteq S_3$. In fact, if $S_1 \subseteq S_2$ and $S_2 \subseteq S_1$, then $S_1 = S_2$. This characteristic is described by saying that $\subseteq$ is *antisymmetric*.

An *equivalence relation* is a relation which is reflexive, symmetric, and transitive. It captures some notion of equivalence, sameness, or similarity. Given an equivalence relation E on a set A and $a \, \varepsilon \, A$, then the set of all elements x of A such that $x \, E \, a$ (x is related to a by E) is written $[a]$ and is called the *equivalence class* of a. An equivalence relation E on a set A partitions A into disjoint subsets called equivalence classes. The collection of equivalence classes (partitions), denoted by A/E, is called the *quotient set* of A by E.

Example: For $n \geq 2$, congruence modulo n on the set of integers is an equivalence relation. ("x is congruent to y modulo n" means that the difference $x - y$ is divisible by n.) If we take $n = 3$, then there are exactly 3 distinct equivalence classes in the quotient set of **Z** by congruence modulo 3.

$[0] = \{0, 3, -3, 6, -6, 9, -9,...\}$

$[1] = \{1, 4, -2, 7, -5, 10, -8,...\}$

$[2] = \{2, 5, -1, 8, -4, 11, -7,...\}$

Note that the equivalence classes are pairwise disjoint and that **Z** = $[0] \cup [1] \cup [2]$.

4.1.1   POSETS AND LATTICES

A binary relation on a set S that is reflexive, antisymmetric, and transitive is called a *partial order* on S. The set S together with a partial order on S is called a *partially ordered set* or *poset*. We usually denote

30

a partial order relation by ≤, and $x \le y$ is read "x precedes y". Two elements in a poset are said to be *comparable* if one precedes the other. The word "partial" used in defining a poset $(S, \le)$ signifies that some elements of S may not be comparable with respect to ≤. However, if every pair of elements in S are comparable with respect to ≤, then S is said to be *totally ordered* or linearly ordered with respect to ≤.

Given a poset $(S, \le)$, if $x \le y$ then either $x = y$ or $x \ne y$. If $x \le y$ but $x \ne y$, we write $x < y$ and say that x is a *predecessor* of y or y is a *successor* of x. A given y may have many predecessors, but if $x < y$ and there is no z such that $x < z < y$, then x is said to be an *immediate predecessor* of y.

We can graph the poset $(S, \le)$ if S is finite. Each of the elements of S is denoted by a dot, called a *node*, or *vertex*, of the graph. If x is an immediate predecessor of y, then the node for y is placed above the node for x, and the two nodes are connected by a straight line segment.

Example: Let $S = \{1, 2, 3, 4, 6, 8, 9, 12, 16, 18, 24\}$ be partially ordered by the divisibility relation "x divides y". The graph for this partially ordered set is shown in Figure 5. (Such graphs are called *Hasse diagrams*.)

Figure 5:

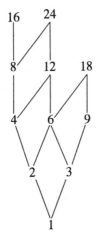

Example. Let S be the set of partitions of the number 5, that is, let $S = \{5, 4 + 1, 3 + 2, 3 + 1 + 1, 2 + 2 + 1, 2 + 1 + 1 + 1, 1 + 1 + 1 + 1 + 1\}$. Let $\leq$ be defined such that $P_1 \leq P_2$ if the integers in partition P_2 can be further subdivided to obtain the integers in the partition P_1. For example, the partition $2 + 2 + 1$ precedes the partition $3 + 2$, while $2 + 2 + 1$ and $3 + 1 + 1$ are non-comparable. Figure 6 gives the Hasse diagram for this poset.

Figure 6:

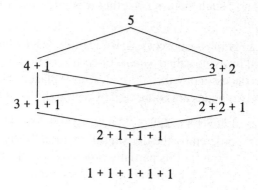

An element x in a poset $(S, \leq)$ is called a *maximal element* if no element succeeds x. Similarly, an element y is called a *minimal element* if no element precedes y. There can be more than one maximal and more than one minimal element. An infinite poset may have neither maximal nor minimal elements.

Let A be a subset of a partially ordered set $(S, \leq)$. An element x is called an *upper bound* of A if x succeeds every element of A. If an upper bound of A precedes every upper bound of A, then it is called the *least upper bound* (*lub*) of A denoted by A. Analogous definitions exist for *lower bound* and the *greatest lower bound* (*glb*) of A denoted by A.

Example: Consider the set of natural numbers N partially ordered by the relation of divisibility. The greatest common divisor of numbers a and b, denoted $gcd(a,b)$, is the largest number which divides a and b. The least common multiple of numbers a and b, denoted $lcm(a,b)$, is the smallest number divisible by both a and b. It can be shown that every common divisor of numbers a and b

divides *lcm(a,b)* and that *gcd(a,b)* divides every multiple of *a* and *b*. Therefore, for any given subset of *N*, the greatest common divisor of the numbers involved will be the greatest lower bound and the least common multiple of the numbers involved will be the least upper bound.

A poset in which every finite subset has a least upper bound and a greatest lower bound is called a *lattice*. A poset in which every subset (not just finite subsets) has a *lub* and a *glb* is called a *complete lattice*. The element of a complete lattice which is the least upper bound of the whole lattice is called "top" and is denoted by $_\top$; the element which is the greatest lower bound of the whole lattice is called "bottom" and is denoted by $_\perp$.

Example: An example of a complete lattice is a power set of a given set *S* with $\subseteq$ as its order relation. Figure 7a gives the diagram of the complete lattice of all subsets of {*a, b, c*}.

An example of a lattice which is not complete is the integers with $\leq$. Every finite set of integers has a maximum and a minimum, but infinite sets of integers (such as positive primes) do not.

Figure 7b, the diagram for a poset in which every subset has a least upper bound and a greatest lower bound. Specifically, for the subset {*d, j, i*} while nodes *b* and *a* represent upper bounds and nodes *f* and *h* represent lower bounds, there is neither a *lub* nor a *glb*. The reader is invited to convince herself that in Figure 7c the dashed line segments indicate relations that must be added to make this poset a lattice.

Figure 7a:

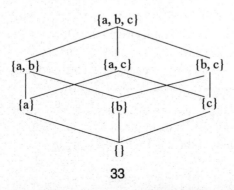

33

Figure 7b:

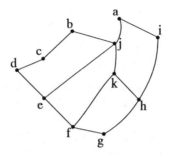

Figure 7c:

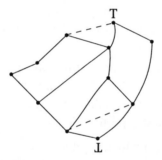

4.2 OPERATIONS ON RELATIONS

Because relations are sets, all of set-theoretic operations such as union and intersection can also be applied to them. In this section, however, we examine a number of operations particular to relations.

The *inverse* of binary relation R from set A to set B is denoted by R^{-1} and is defined by: $b\,R^{-1}\,a$ if and only if $a\,R\,b$. It is not hard to see that the domain of R^{-1} is the range of R and that the range of R^{-1} is the domain of R. Note that the relation R^{-1} consists of all ordered pairs in R written in reverse order. Thus, it should be clear that $(R^{-1})^{-1} = R$.

The *transitive closure* of a binary relation R is the smallest transitive relation containing R. For example, if R is the relation $\{(1, 2), (2, 3), (3, 4), (2,1)\}$ defined on the set $\{1, 2, 3, 4\}$, its transitive closure can be computed by determining, for each element of the set, all ordered pairs whose first component is that element and that must be added to R in order to conform to transitivity. Starting with element 1, we find that we must add $(1, 1)$ (since $1\ R\ 2$ and $2\ R\ 1$), $(1, 3)$ (since $1\ R\ 2$ and $2\ R\ 3$), and $(1, 4)$ (since $1\ R\ 2$, $2\ R\ 3$, and $3\ R\ 4$). Starting with element 2, we determine that we must add $(2, 2)$ (since $2\ R\ 1$ and $1\ R\ 2$), and $(2, 4)$ (since $2\ R\ 3$ and $3\ R\ 4$). Starting with element 3 or element 4, we find that no additional ordered pairs need to be added. Therefore, the transitive closure of R would be $\{(1, 1), (1, 2), (1, 3), (1, 4), (2, 1), (2, 2), (2, 3), (2, 4), (3, 4)\}$.

Let A, B, and C be sets. Let R be a relation from A to B and S a relation from B to C. The *composition* of R and S, written $R \circ S$, is a relation from A to C defined as follows. If a is in A and c is in C, then $a\ (R \circ S)$ c if and only if there exists some b in B such that $a\ R\ b$ and $b\ S\ c$. A good way to compute $R \circ S$ is to employ the *matrix representation* of relations involved. This is illustrated in the following example.

Example: Let $A = \{a, b, c, d\}$, $B = \{1, 2, 3, 4\}$, $C = \{x, y, z\}$, and let $R = \{(a,1), (b, 4), (c, 1), (c, 2), (c, 4)\}$ and $S = \{(2, x), (2, z), (3, y), (4, z)\}$.

The matrix of the relation R is obtained by forming a rectangular array whose rows are labeled by the elements of A and whose columns are labeled by the elements of B. A 1 or 0 is placed in row a, column 1 according to whether or not $a\ R\ 1$. The matrices for R and S are as follows:

$$M_R = \begin{pmatrix} 1 & 0 & 0 & 0 \\ 0 & 0 & 0 & 1 \\ 1 & 1 & 0 & 1 \\ 0 & 0 & 0 & 0 \end{pmatrix} \qquad M_S = \begin{pmatrix} 0 & 0 & 0 \\ 1 & 0 & 1 \\ 0 & 1 & 0 \\ 0 & 0 & 1 \end{pmatrix}$$

Multiplying M_R and M_S we obtain the matrix:

$$\begin{matrix} 0 & 0 & 0 \\ 0 & 0 & 1 \\ 1 & 0 & 2 \\ 0 & 0 & 0 \end{matrix}$$

The nonzero entries in this matrix indicate which elements are related by $R \circ S$. Thus $M = M_R \times M_S$ and $M_{R \times S}$ have the same nonzero entries. Therefore, $R \circ S = \{(b, z), (c, x), (c, z)\}$.

4.3   MAPPINGS AND FUNCTIONS

The term *mapping* relates to the concept of assigning to an element in one set a *unique* element in another set. Consider the situations depicted in Figure 8a. While F_1 is a perfectly good relation from A to B, it has the property of assigning two different elements of B, i.e., 1 and 2, to the same element in A, i.e., x. The relations F_2 and F_3 do not have this property and associate with each element of A a single element in B. Thus, F_2 and F_3 are functions, whereas F_1 is not. Now consider the *inverse* relations F_1^{-1}, F_2^{-1}, and F_3^{-1}. These are represented by Figure 8b. Note that although F_1 is not a function, F_1^{-1} is; and, even though F_2 is a function, F_2^{-1} is not. Only in the third case, i.e., F_3 and F_3^{-1}, both a relation and its inverse are functions.

Figure 8a:

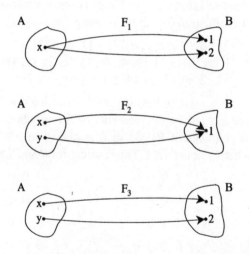

Figure 8b:

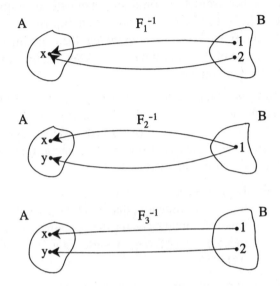

A *function* or mapping f from a set A to a set B is a 'rule' or 'method' which pairs elements of the set A with unique elements of the set B. We write, $f: A \to B$, to indicate that f is a mapping from the set A to the set B, and write $f(a) = b$ to indicate that f maps the value a *into* value b. B is called the *codomain* of the mapping f, while A is sometimes called its domain. The *domain* of f is that subset of A for which f gives a value in B, and the *range* of f is that subset of B to which f maps a value. A function f is called *total* if its domain is the entire set A, otherwise it is called a *partial* function.

A mapping $M: A \to B$ is said to be from A *onto* B or a *surjection*, if for every b in B there is an a in A such that $M(a) = b$. In other words, the range of M is the entire set B. A mapping $M: A \to B$ is *one-to-one* or an *injection* if for each b in B there is at most one a in A such that $M(a) = b$. A total, one-to-one mapping from A onto B is called a *bijection* or *one-to-one correspondence*. In this case, there is a one-to-one pairing of the members of A and B, and the two sets are sometimes called *isomorphic* which means "having the same shape."

The concept of a bijection allows us to provide a somewhat more formal definition of *cardinality*. Specifically, two sets A and B have the same cardinality if there is a bijection from A onto B. A set A is of cardinality n if there is a bijection from A onto $\{1, 2, ..., n\}$. A set A is *finite* if there is a bijection from A to a finite subset of natural numbers which is definable as $\{x \mid x \in N \wedge x < n\}$ for some natural number n. A set A is *infinite* if it is not finite. A set A is *countable* if there is a bijection from A onto some subset of the natural numbers. A set A is *uncountable* if it is infinite and not countable.

In the same way that relations can be composed, so can functions. Let $f: A \to B$, and $g: B \to C$. Then the *composition function*, $g \circ f$, is a function from A to C defined by $(g \circ f)(a) = g(f(a))$. Note that the function $g \circ f$ is applied from right to left; that is, function f is first applied to a and then function g is applied to the result.

Example: Let $f: R \to R$ be defined by $f(x) = x^2$. Let $g: R \to R$ be the *ceiling* function, that is, $g(x) = \lceil x \rceil$ which gives the smallest integer greater than x. Now, $(g \circ f)(3.5) = g(f(3.5)) = g(12.25) = 13$ while $(f \circ g)(3.5) = f(g(3.5)) = f(4) = 16$.

It should be apparent that the composition of a function $f: A \to B$ and its *inverse* $f^{-1}: B \to A$, *if it exists*, will map each element of A onto itself, that is, $f \circ f^{-1} = I_A$ (the identity function on A). For bijective functions, the inverse function must exist. In fact, f is a bijection if and only if f^{-1} exists.

We conclude this section by noting that the definition of mapping allows functions of more than one variable. Specifically, we can have a function $f: A_1 \times A_2 \times ... \times A_n \to B$ that associates with each ordered n-tuple $<a_1, a_2, ..., a_n>$, $a_i \in A_i$, a unique element of B.

CHAPTER 5

ALGEBRAIC *STRUCTURES* AND HOMOMORPHISMS

5.1　OPERATIONS

A total mapping $M: A \times A \to A$ is called *a dyadic operation*. The term 'dyadic' indicates that the operation takes two arguments. In general, the *adicity* or *arity* of an operation is the number of arguments it takes. Thus, in arithmetic we can regard '+', i.e., addition, as a mapping which, given any pair of elements, produces a new value.

The operations with which we are most familiar are usually written in *infix notation*, where the 'name' of the operator is placed between its arguments, e.g., $2 + 3$. However, the infix notation becomes impractical for operations involving more than two arguments, which are written by having the 'name' of the operation precede its arguments, e.g., $f(a, b, c)$. In fact, the more familiar operations can also be expressed in this general way which is known as *Polish notation*. Thus, $+ab$ means $(a + b)$ in infix notation and $*+ ab - cd$ means $(a + b)*(c - d)$. Note that, Polish notation permits us to avoid the use of parentheses.

An *identity element* is associated with some dyadic operations. Given the operation $\square: A \times A \to A$, the element i in A is an identity element for $\square$ if $\forall a \ \varepsilon A, a \square i = a = i \square a$. For example, the identity element for $+: Z \times Z \to Z$ is 0, since for all integers $i, i + 0 = i = 0 + i$, and the identity element for $*: Z \times Z \to Z$ is 1, since for all integers $i, i*1 = i = 1*i$.

An element $a \ \varepsilon A$ has an *inverse b* under an operation $\square$ if $a \square b =$

$b \square a = i$, where i is the identity element. For example, every element of the integers has an inverse with respect to the addition operation. And, every element of the reals except 0 has an inverse with respect to the multiplication operation. By convention, for any operation named "+", the identity element is named "0" and the inverse of "a" is written "$-a$", and for any operation named "*", the identity element is named "1" and the inverse of "a" is written "a^{-1}".

An operation $\square: A \times A \to A$ is *associative* if $\forall a, b, c \in A$, $a \square (b \square c) = (a \square b) \square c$. For example, the string concatenation operation, $|$, is assoc-iative, since given string S_1, S_2, and S_3 over an alphabet A, $S_1 \mid (S_2 \mid S_3) = (S_1 \mid S_2) \mid S_3$.

A dyadic operation $\square: A \times A \to A$ is said to be *commutative* or synonymously *Abelian*, if $\forall a, b \in A$, $a \square b = b \square a$.

An operation $\square: A \times A \to A$ is said to be *closed* over A if $\forall (a_1, a_2) \in A \times A$, $a_1 \square a_2$ is defined and is a member of A.

5.2 GROUPS, RINGS AND FIELDS, AND ALGEBRAS

A set with a total associative operation is called a *semigroup*. A semigroup with an identity element is called a *monoid*. A monoid in which every element has an inverse is called a *group*. A semigroup, monoid, or group whose operation is commutative is called an Abelian semigroup, monoid, or group.

Example: The integers or reals with + are groups; with * they are monoids. The identity for * is 1, but no integer except 1 and -1 has an inverse, and the real 0 has no inverse under *.

The Booleans with ^ is a monoid: ^ is associative, True is an identity, but False has no inverse.

The Booleans with $\to$ is not a semigroup since $\to$ is not associative.

A *ring* is a set A with two operations +, *, such that *i*) A is a group with respect to +, *ii*) A is closed under *, *iii*) * is associative, and *iv*) * is *distributive* with respect to +, that is, $\forall a, b, c \in A$, $a*(b + c) = (a * b) + (a * c)$ and $(a + b)*c = (a * c) + (b * c)$.

A ring which is a group with respect to *, except that 0 has no inverse under *, is called *a field*.

Example: The integers with respect to + and * form a ring. The integers with respect to + and / are not a ring since / is not associative. The reals with respect to + and * form a field. The Booleans with respect to ^ and V are not a ring since the Booleans are not a group with respect to ^. The integers modulo n with respect to + and * form a field if n is prime, otherwise a ring.

A set with a finite set of operations of various adicities is called an *algebra*. Semigroups, monoids, groups, rings, and fields are all particular kinds of algebras. A finite collection of sets closed under set intersection (∩) and set union (∪) is an algebra. The set of all strings over an alphabet with the operation of concatenation is an algebra.

5.2.1 BOOLEAN ALGEBRA

A Boolean algebra is an algebra with three operations called addition or sum (+), multiplication or product (*), and complementation or complement (,). Sum and product operators are both associative and commutative, and we have both an additive identity denoted by 0 and a multiplicative identity denoted by 1. Also, multiplication distributes over addition, and furthermore, addition distributes over multiplication (i.e., $a + (b * c) = (a + b) * (a + c)$). The complementation operator defines the complement of an element a to be the element a' such that $a + a' = 1$, and $a * a' = 0$.

A few important properties that hold for all elements of a Boolean algebra are listed below.

i) $a + a = a$ and $a * a = a$ ....Idempotency

ii) $a + 1 = 1$ and $a * 0 = 0$ ....Boundedness

iii) $a + (a * b) = a$ and $a * (a + b) = a$ ....Absorption

iv) $(a + b) + c = a + (b + c)$ and
     $(a * b) * c = a * (b * c)$ ....Associativity

v) $(a')' = a$ ....Involution Law

vi) $0' = 1$ and $1' = 0$ ....Identity Complement Laws

vii) $(a + b)' = a' * b'$ and $(a * b)' = a' + b'$ ......De Morgan's Laws

The *principle of duality* states that if a property can be established for a Boolean algebra, then the property derived from the given one by the interchange of the operations of sum and product, and of the elements 0 and 1, holds and can be established by corresponding changes in the proof of the given property.

Example: Given two elements a and b of a Boolean algebra, we can prove that $a + a'b = a + b$. (As is customary, we denote multiplication, $a*b$, by juxtaposition, ab.)

$a + a'b = a1 + a'b = a(1 + b) + a'b = a + ab + a'b = a + (a + a')b = a + 1b = a + b$.

By applying the principle of duality we can show that: $a\,(a' + b) = ab$.

It can be shown that with the exception of the one-element Boolean algebra, any finite Boolean algebra has an even number of elements. The addition and multiplication tables for the *eight-element* Boolean algebra are shown below:

+	0	a	a'	b	b'	c	c'	1
0	0	a	a'	b	b'	c	c'	1
a	a	a	1	c'	b'	b'	c'	1
a'	a'	1	a'	a'	1	a'	1	1
b	b	c'	a'	b	1	a'	c'	1
b'	b'	b'	1	1	b'	b'	1	1
c	c	b'	a'	a'	b'	c	1	1
c'	c'	c'	1	c°	1	1	c'	1
1	1	1	1	1	1	1	1	1

*	0	a	a'	b	b'	c	c'	1
0	0	0	0	0	0	0	0	0
a	0	a	0	0	a	0	a	a
a'	0	0	a'	b	c	c	b	a'
b	0	0	b	b	0	0	b	b
b'	0	a	c	0	b'	c	a	b'
c	0	0	c	0	c	c	0	c
c'	0	a	b	b	a	0	c'	c'
1	0	a	a'	b	b'	c	c'	1

Moreover, it can be established that each finite Boolean algebra is *isomorphic* to an algebra consisting of the power set of a finite set together with operations of set union, set intersection, and set complement. And, since the number of elements in the power set of a set with n elements is $2n$, we may conclude that any finite Boolean algebra has 2^n elements for some positive integer n.

Example: The eight-element Boolean algebra is isomorphic to the power set of the set $\{a, b, c\}$ as the following mapping demonstrates.

0	$\{\}$	a	$\{a\}$
a'	$\{b, c\}$	b	$\{b\}$
b'	$\{a, c\}$	c	$\{c\}$
c'	$\{a, b\}$	1	$\{a, b, c\}$

A *Boolean function* is any function on a Boolean algebra that can be derived from the *constant function* and the *projection functions* by the use of the Boolean operations. Let $a_1, a_2, ..., a_n$ be variables defined over a Boolean algebra, and let a be a fixed element of the algebra. Then, $f(a_1, a_2, ..., a_n) = a$ is the constant function and $g_i(a_1, a_2, ..., a_n) = a_i$ is the ith projection function. Furthermore, let f and g be any two Boolean functions of n variables, then $(f(a_1, a_2, ..., a_n)$, and $f(a_1, a_2, ..., a_n) + g(a_1, a_2, ..., a_n)$ and $f(a_1, a_2, ..., a_n) * g(a_1, a_2, ..., a_n)$ are also Boolean functions of n variables. Iteration of this process a finite number of times will result in the definition of the complete set of Boolean functions on n variables.

Informally, the Boolean functions may be written as *Boolean expressions* such as: $a + a'b + c$ or $(a + b'c)((ab'c' + b)' + a'c)$. (As is customary, we denote multiplication, $a \times b$, by juxtaposition, ab.) Because of De Morgan's laws, involution, absorption, and other properties of Boolean algebras, there is no unique expression for a Boolean function. However, we can define a standard or *canonical form* to which all Boolean functions may be transformed. The *disjunctive normal form* is such a standard form.

A Boolean expression is said to be in disjunctive normal form if it is a *fundamental product* or the sum of two or more fundamental products of which none is *included* in another. Let us define a *literal* to be a variable or complemented variable, for example, a, a', b, and b' are literals while $(ab')'$ is not. By a fundamental product we mean a literal or product of literals in which no two literals involve the same variable. For example, ab, $a'b$, $ab'c$, and bc' are all examples of fundamental products, but $aba'c$ is not since the literals a and a' involve the same variable. We will say a fundamental product is included in another if all literals of the former are included in the literals of the latter, e.g.. $b'c$ is included in $ab'c$.

Example: The boolean expression $a(b'c)'$ can be transformed to disjunctive normal form as follows:

$$a(b'\ c)' = a(b + c') = ab + ac'.$$

Similarly, the expression $(a' + b)c + b'$ may be transformed as follows: $(a' + b)c + b' = a'c + bc + b'$.

5.3   HOMOMORPHISMS

If an operation '+' is defined on two distinct sets A and B, then a mapping $h: A \rightarrow B$ is a *homomorphism* if $\forall a_1, a_2 \in A$, $h(a_1 + a_2) = h(a_1) + h(a_2)$. Homomorphisms may exist between two semigroups, two monoids, two groups, two rings, two fields, as well as other algebraic structures. For rings and fields which have two operations + and *, we would require that for all a_1 and a_2, $h(a_1 + a_2) = h(a_1) + h(a_2)$ and $h(a_1 * a_2) = h(a_1) * h(a_2)$.

A homomorphism will preserve various properties of an algebra. For example, if A and B are both groups, h will map identities to identities (i.e., $h(0)$ is an identity for B) and inverses to inverses (i.e., $h(-a) = -h(a)$). Rules, like associativity and commutativity, are preserved by homomorphisms (e.g., $h(a_1) + (h(a_2) + h(a_3)) = (h(a_1) + h(a_2)) + h(a_3)$). And if h maps a ring A to ring B, then the distributive law is also preserved.

If a homomorphism $h: A \rightarrow B$ maps A *onto* B, that is, if every element of B is an *image* of some element of A under h, then h is said to be an *epimorphism*. A homomorphism which is an *injection*, that is, one-to-one, is called a *monomorphism*. A homomorphism which is both an epimorphism and a monomorphism is called an *isomorphism*. If there exists an isomorphism between two algebraic structures, then they are *essentially the same*. The only difference between them is the names of their elements.

Example: Consider the set $\{\{\}, \{a\}\}$ with the operations $\cup$ (set union) and $\cap$ (set intersection) as expressed below:

$\cup$	$\{\}$	$\{a\}$
$\{\}$	$\{\}$	$\{a\}$
$\{a\}$	$\{a\}$	$\{a\}$

$\cap$	$\{\}$	$\{a\}$
$\{\}$	$\{\}$	$\{\}$
$\{a\}$	$\{\}$	$\{a\}$

Now consider the familiar Boolean algebra $\{T, F\}$ with the operations $\wedge$ and $\vee$:

^	T	F		$\vee$	T	F
T	T	F		T	T	T
F	F	F		F	T	F

A homomorphism h mapping $T \to \{\}$ and $F \to \{a\}$ preserves the operations: ^ is mapped to $\cup$ and $\vee$ to $\cap$. Since h is one-to-one and onto, it is an isomorphism.

In the same manner that bijections have inverses, so do isomorphisms. In this case the mapping h^{-1} mapping $\{\} \to T$ and $\{a\} \to F$ will map $\cup$ to ^ and $\cap$ to $\vee$.

CHAPTER 6

GRAPHS AND TREES

6.1 GRAPH TERMINOLOGY

Of all mathematical structures, graphs are probably the most widely used. A graph $G = <V, E>$ is an ordered pair of finite sets V and E. The elements of V are called *vertices* or *nodes*. The elements of E are called *edges* or *arcs*. Each edge in E joins two distinct vertices in V. Figure 9 gives some examples of graphs where vertices are represented by circles and edges by lines.

Figure 9:

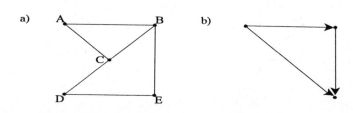

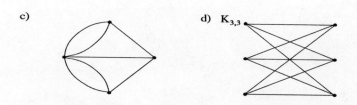

e) K_5

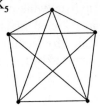

f)

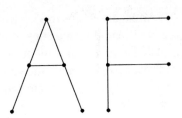

g)

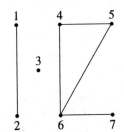

h)

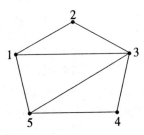

i)

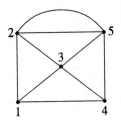

j)

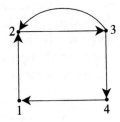

k)

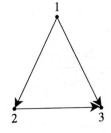

l)

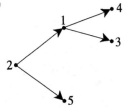

An edge with an arrow head is called a *directed edge* while one without is called an undirected edge. If all the edges are directed then the graph is called a *directed graph* or *digraph* (see Figure 9b). Notationally, an undirected edge connecting vertices i and j is denoted by (i, j) or equivalently (j, i), while a directed edge *from i to j* is denoted by $<i, j>$ and is distinguished from $<j, i>$.

By definition, a graph does not contain multiple copies of the same edge. For an undirected graph this means that there can be at most one edge between any pair of vertices. In the case of a directed graph, this means that there can be at most one edge from vertex i to vertex j and one from j to i. Also, we require that a graph contain no *self edges* (or *loops*), that is, no edges connecting a vertex to itself. A *multigraph* (see Figure 9c) allows multiple edges but no loops, while a *pseudograph* permits both loops and multiple edges.

A graph $G = <V, E>$ is said to be *bipartite* if its vertices V can be partitioned into two subsets M and N such that each edge in G connects a vertex in M to a vertex in N (see Figure 9d). By a *complete bipartite* graph we mean that each vertex in M is connected to each vertex in N and denote such a graph by K_{mn} where $m = |M|$ and $n = |N|$.

Let $G = <V, E>$ be an undirected graph. If $|V| = 0$ then the graph is said to be *empty*. Vertices i and j are *adjacent* in G if and only if the edge (i, j) is in E. The edge (i, j) is said to be *incident* on the vertices i and j. The *degree* d_i of vertex i is the number of edges incident on vertex i. Since each edge in an undirected graph is incident on exactly two vertices, the sum of the degrees of the vertices equals two times the number of edges.

Example: A house has a single entrance. Each room has a certain number of doors leading to adjacent rooms. There is a ghost in each room that has an even number of doors. A) Prove that there is a room without a ghost. B) Prove that a person entering the house from outside can always reach the room where there is no ghost without going through any door twice.

A) Represent the rooms and the outside by vertices. Represent the doors by edges. Then the vertex representing the outside has a degree of one. Since the sum of the degrees of all vertices must be an even number, there must be another vertex with an odd degree. That

48

room represents a room without a ghost.

B) If we start at the vertex representing the outside and walk along the edges, never repeating an edge, we will eventually come to a vertex with an odd degree, since if a vertex has an even degree, every time we visit it there is an edge left to leave by. The odd vertex represents a room with an odd number of doors.

An undirected graph is said to be *complete* if each vertex is connected to every other vertex (see Figure 9e). The complete graph with n vertices is denoted by K_n and contains exactly $n(n-1)/2$ edges.

The undirected graph $G' = <V', E'>$ is a *subgraph* of the graph $G = <V, E>$ if and only if $V' \subseteq V$ and $E' \subseteq E$. From the definition of the complete graph it follows that every n vertex undirected graph is a subgraph of K_n. Since there are $n(n-1)/2$ possible different edges in an n vertex graph and each of which may or may not be included in a particular graph, there are $2^{n(n-1)/2}$ different n vertex graphs.

Example: What is the total number of subgraphs of K_n?

Let us count the number of subgraphs with 0, 1, 2, ..., and n vertices. There is only one subgraph of K_n with 0 vertices – the empty graph. For any i ($\leq n$), the number of possible graphs is $2_{i(i-1)/2}$, and since there are $C(n,i)$ ways of selecting i vertices out of n, the number of subgraphs of K_n having exactly i vertices is: $C(n,i) \circ 2^{i(i-1)/2}$. Therefore, the total number of subgraphs for K_n is given by: $\Sigma_{i=0,n} C(n,i) \circ 2^{i(i-1)/2}$.

Two graphs $G_1 = <V_1, E_1>$ and $G_2 = <V_2, E_2>$ are said to be *isomorphic* if and only if a bijective mapping $f: V_1 \rightarrow V_2$ can be defined such that (i, j) is an edge in G_1 if and only if $(f(i), f(j))$ is an edge in G_2 (see Figure 9f).

A sequence of vertices $P = i_1, i_2, ..., i_k$ is an i_1 to i_k *path* in the graph $G = <V, E>$ if and only if the edge (i_j, i_{j+1}) is in E for every j, $1 \leq j \leq k$. An edge (u, v) is *on the path* if and only if there exists a j such that $u = i_j$ and $v = i_{j+1}$ or $u = i_j$ and $v = i_{j-1}$. The *length* of a path is the number of edges on it. The path P is a *simple path* if and only if all vertices (except possibly the first and the last) and all edges on it are distinct. A *cycle* is a simple path in which the first and last vertices are the same.

An undirected graph $G = <V, E>$ is *connected* if and only if for every pair of vertices i and j in V, $i \neq j$, there is an i to j path in G. Figure 9g shows a graph that is not connected and has three *connected components*.

A path is said to be a *Euler path* if and only if it uses each edge in E exactly once. A Euler path in which the first and the last vertices are the same is called a *Euler circuit*. A simple characterization of graphs that contain Euler paths is given by the following theorem. Let $G = <V, E>$ be a connected undirected graph. G has a Euler path if and only if it contains either zero or exactly two vertices with an odd degree. G contains an Euler circuit if and only if all vertices have an even degree. In Figure 9h, the paths 1, 2, 3, 1, 5, 3, 4, 5 and 1, 2, 3, 4, 5, 1, 3, 5 are Euler paths.

Example: The general idea of paths and specifically Euler paths may be extended to multigraphs. In fact, the preceding theorem may be generalized by stating that any finite connected graph with two odd vertices is *traversable*. Figure 9c is an example of a multigraph that is not traversable, while Figure 9i is a multigraph that can be traversed as follows: <1, 2> <2, 3> <3, 1> <1, 4> <4, 3> <3, 5> <5, 2> <2, 5> <5, 4>.

An *acyclic* graph is an undirected graph that contains no cycles. A graph is a *tree* if and only if it is both connected and acyclic. Since each connected component of an acyclic graph is a tree, an acyclic graph is also called a *forest*.

Most of the terminology for undirected graphs extends in an obvious way to digraphs. The directed edge $<i, j>$ is said to be *incident from* i and *incident to* j. If an edge $<i, j>$ is in E, then vertex i is *adjacent to* vertex j and vertex j is *adjacent from* vertex i. The *in-degree* of a vertex is the number of edges incident to it. The *out-degree* of a vertex is the number of edges incident from it. Since each edge is incident to exactly one vertex and incident from exactly one vertex, it follows that the sum of in-degrees for all vertices is the same as the sum of out-degrees for all vertices and is equal to the number of edges in the digraph. A *complete* digraph of n vertices has exactly $n(n-1)$ directed edges.

Let $G = <V, E>$ be a directed graph. A sequence of vertices $P = i_1, i_2, ..., i_k$ is a *directed path* from i_1 to i_k if and only if the edge $<i_j, i_{j+1}>$ is in E for every j, $1 \leq j \leq k$. The *length* of a directed path is the number of edges on it. The path P is a *simple directed path* if and only if all vertices (except possibly the first and the last) are distinct. A *directed cycle* is a simple directed path in which the first and last vertices are the same.

A sequence of vertices $P = i_1, i_2, ..., i_k$ is a *semipath* from i_1 to i_k if and only if either edge $<i_j, i_{j+1}>$ or edge $<i_{j+1}, i_j>$ is in E for every j, $1 \le j \le k$.

A digraph is *strongly connected* if and only if it contains a directed path from i to j and from j to i for every pair of distinct vertices i and j (see Figure 9j). A digraph is *weakly connected* if and only if it contains an i to j semipath for every pair of distinct vertices i and j.

A *directed acyclic graph* (*dag*) is a digraph that contains no directed cycles (see Figure 9k). No strongly connected digraph with more than one vertex can be a *dag*.

A *directed tree* is a weakly connected digraph that contains no *semicycles*. A *rooted tree* is a directed tree in which exactly one vertex (called the *root*) has an in-degree of 0 and the remaining vertices have an in-degree of 1 (see Figure 9l). An n vertex directed tree contains exactly $n-1$ edges.

A graph or multigraph which can be drawn in the plane in such a way that no two of its edges intersect is said to be *planar*. Kuratowski's theorem states that a graph is non-planar if and only if it contains a subgraph *homemorphic* to $K_{3,3}$ (see Figure 9d) or K_5 (see Figure 9e). Two graphs G_1 and G_2 are homemorphic if isomorphic graphs can be obtained by inserting new vertices of degree 2 on the edges of G_1 and/or G_2 (see Figure 10 for an example).

Figure 10:

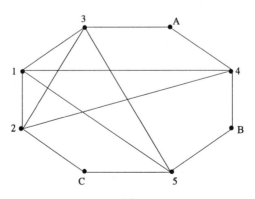

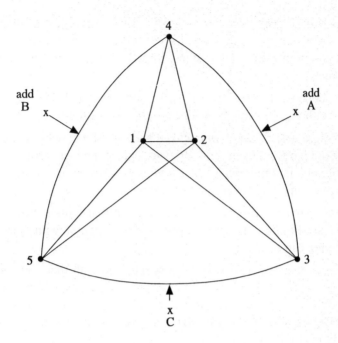

6.2   COMPUTER REPRESENTATION OF GRAPHS

Let $G = <V, E>$ be a finite graph (not necessarily connected), without multiple edges or loops . Let $V = \{v_1, v_2, ..., v_m\}$ and $E = \{e_1, e_2, ..., e_n\}$. An *incidence matrix* for G is an $m \times n$ matrix M such that $m_{ij} = 1$ if edge e_j is *incident from* vertex v_i, and -1 if e_j is *incident to* v_i, and 0 otherwise. A (vertex) *adjacency matrix* for G is an $m \times m$ matrix A such that $a_{ij} = 1$ if there is an edge *from* v_i *to* v_j. If G is an undirected graph, then the adjacency matrix will be symmetric.

Example: The incidence and adjacency matrices for the directed graph shown in Figure 11 are given below.

	e_1	e_2	e_3	e_4	e_5
v_1	1	0	0	0	0
v_2	-1	1	-1	1	0
v_3	0	-1	1	0	-1
v_4	0	0	0	-1	1

Incidence Matrix

	v_1	v_2	v_3	v_4
v_1	0	1	0	0
v_2	0	0	1	1
v_3	0	1	0	0
v_4	0	0	1	0

Adjacency Matrix

Figure 11:

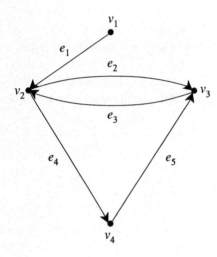

Since every entry in the adjacency matrix is either 0 or 1, it takes n^2 bits to store the adjacency matrix of an n vertex graph in a computer. For undirected graphs the space requirements can be reduced to $(n^2 - n)/2$ bits since the matrix is symmetric and only the elements above or below the main diagonal need to be stored explicitly.

The adjacency matrix representation also affords us some nice computational properties. For example, it can be shown that A^n gives the number of paths of length n from one node to another.

6.3   GRAPH ALGORITHMS

Let $G = <V, E>$ be a graph (either directed or undirected). We say vertex j is *reachable* from vertex i if and only if there is a (directed) path from i to j. Consider the digraph in Figure 12. One way to determine all nodes reachable from node A is to first determine the set of vertices adjacent from A. This set is $\{B, F\}$. Next, we determine the set of new vertices (i.e., vertices not yet reached) adjacent from vertices in $\{B, F\}$. This set is $\{C, G, E, I\}$. The set of new vertices adjacent from a vertex in $\{C, G, E, I\}$ is $\{H\}$. There are no new vertices adjacent from H. Therefore, $\{A, B, F, C, G, E, I, H\}$ is the set of vertices reachable from A. This method is called *breadth first search*.

In the *depth first search* algorithm we start at node A and mark it as reached. Next, we select an unreached vertex adjacent from A. If such a vertex does not exist, the search terminates. Otherwise, a depth first search is *recursively* initiated at the new node. In the case of our example, the depth first search would visit the nodes in the following fashion: $A \rightarrow B \rightarrow C \rightarrow H \rightarrow G \rightarrow \{H\} \rightarrow I \rightarrow E \rightarrow \{I\} \rightarrow \{H\} \rightarrow \{C\} \rightarrow \{B\} \rightarrow \{A\} \rightarrow F \rightarrow \{A\}$, where the nodes in brackets signify a re-visit as each recursive invocation of the algorithm terminates.

Figure 12:

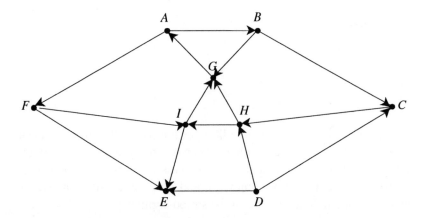

Let $G = <V, E>$ be an undirected graph. A subgraph $G' = <V', E'>$ of G is a *spanning tree* of G if and only if $V' = V$ and G' is a tree. The edges that lead to new, i.e., unreached, vertices during a breadth first (depth first) search of a graph form a rooted tree called a *breadth first* (depth first) *spanning tree*. Figure 13 gives a graph and a breadth first spanning tree as well as a depth first spanning tree for it.

Figure 13a:

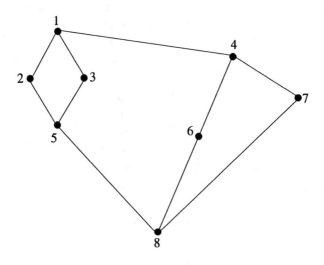

Figure 13b: Breadth First Spanning Tree

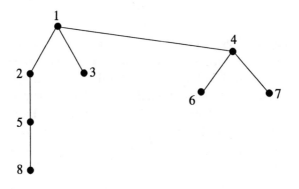

Figure 13c: Depth First Spanning Tree

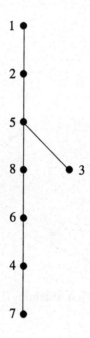

A *weighted* graph is a graph with weights associated with its edges. Figure 14a gives an example of such a graph. The *minimal spanning tree* problem is concerned with finding *one* of the minimum weight (cost) spanning trees of a given weighted graph. An efficient method for solving this problem is known as Kruskal's algorithm. Essentially, the algorithm begins by selecting an edge with least cost (weight). Then, from the remaining edges another least cost edge is selected provided that its inclusion does not create a cycle in the spanning tree being created. When $n - 1$ (n is the number of nodes in the graph) edges have thus been selected, a minimal spanning tree has been derived. Applying the algorithm to the weighted graph in Figure 14a, the selection of edges starts as follows: (E, D), (I, G), (A, B), (D, C), (H, C), (I, E). Next, the edge (G, H) whose weight is 3 is considered. However, its inclusion would create a cycle(G, I, E, D, C, H, G) and so it is ignored. The edge (A, G) is next added. Neither of the two edges with weight 4, i.e., (B, C) and (I, H), can be added without introducing a cycle. The next candidate, edge (I, F) with a weight of 5, is the eighth and final edge to be added. The minimal cost spanning tree is shown in Figure 14b.

Figure 14a:

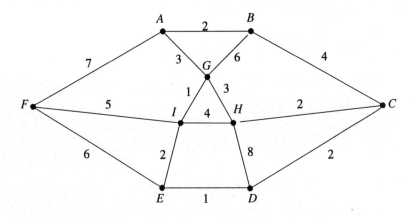

Figure 14b:

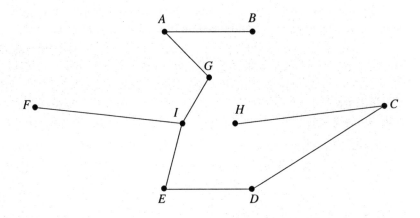

CHAPTER 7

AUTOMATA THEORY

7.1   FINITE STATE MACHINES

A finite state machine (FSM) is an abstract computation model consisting of a finite number of states, a finite number of input symbols, and a finite number of output symbols. Input, output, and state values are only defined for integral values of time. The state of an FSM at time t and the input symbol applied at time t together determine the output symbol produced at time t and the state of the machine at time $t + 1$.

Formally, $M = <S, I, O, f_s, f_o>$ is a finite state machine if S is a finite set of *states*, I is a finite set of input symbols (the *input alphabet*), O is a finite set of output symbols (the *output alphabet*), $f_s: S \times I \rightarrow S$ is the *state transition function*, and $f_o: S \rightarrow O$ is the *output function*. The machine is always initialized to begin in a fixed starting state $s_0 \,\varepsilon\, S$.

Example: Let $S = \{s_0, s_1, s_2, s_3\}, I = \{0, 1\}$, and $O = \{0, 1\}$. Let the functions f_s and f_o be defined in terms of the following *state transition table*.

Present State	Next State Present Input 0	1	Output
s_0	s_0	s_1	0
s_1	s_0	s_2	0
s_2	s_0	s_3	0
s_3	s_0	s_3	1

The table indicates that, for example, $f_s(s_0,1) = s_1$ and $f_o(s_0) = 1$.

We can determine how the machine will perform when presented with the input sequence 0111101011 as follows. Note that the initial 0 in the output string is spurious as it merely reflects the starting state.

Time	t_0	t_1	t_2	t_3	t_4	t_5	t_6	t_7	t_8	t_9
Input	0	1	1	1	1	0	1	0	1	1
State	s_0	s_1	s_2	s_3	s_3	s_0	s_1	s_0	s_1	s_1
Output	0	0	0	1	1	0	0	0	0	0

In a similar fashion, we can show that the input sequence 11011101 produces the output sequence 00000100.

Another way to define this finite state machine is by a labeled directed graph called a *state graph* or *state diagram*. Each state of the FSM with its corresponding output is the label of a node in the graph. f_s, the state transition function is given by directed arcs of the graph, with each arc showing the particular input symbols that produce that particular state transition. Figure 15 gives the state diagram for our example FSM.

Figure 15:

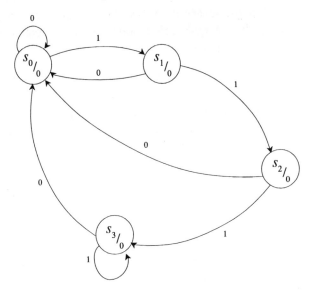

Example: Consider the problem of constructing an FSM to add two binary numbers. The input alphabet consists of pairs of binary digits of the form 00, 01, 10, or 11. These represent the digits of the two numbers which will be presented to the machine from right to left. It should be clear that the machine must account for the following possibilities when presented with a pair of binary digits: 1) the output should be 0 with no carry; 2) the output should be 0 with a carry to the next column; 3) the output should be 1 with no carry; and 4) the output should be 1 with a carry to the next column (e.g., 11 + 11 = 110). Let each of the four possibilities be captured by a state in our FSM. The corresponding outputs are thus: $s_0/0$, $s_1/0$, $s_2/1$, and $s_3/1$. The state transition function can now be determined by considering all possibilities. For example, what should be the next state if the machine is in state s_1 and is presented with 01? Since s_1 reflects the situation when a carry from the previous column exists, the two input digits 0 and 1 must be added to an existing carry resulting in an *output of 0 with a carry to the next column*. This means that the next state must also be s_1. In this manner, the complete state graph may be determined which is shown in Figure 16.

Figure 16:

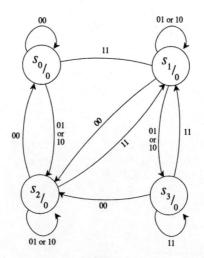

There are two ways of thinking about the role of finite state machines. One way is to view an FSM as a device for *transforming* input sequences into output sequences. The other is to view it as a device for *classifying* input sequences, or deciding which sequences belong to some predesignated set. In the latter role, we will say an FSM *accepts* or *recognizes* a finite input sequence if, in conjunction with the last input symbol, the machine produces a certain predesignated output symbol. For example, the FSM in the first example above recognizes an input pattern consisting of three consecutive 1's by producing an output of 1 after the third 1 in the pattern has been input.

What properties must a set of input sequences have if it is to be *recognizable* by some finite state machine? The answer to this question is provided in the next chapter. Here, it suffices to indicate that no finite state machine can be constructed to recognize input sequences that contain exactly as many 0's as 1's.

7.2   PUSH-DOWN AND LINEAR BOUNDED AUTOMATA

Finite state machines are special-purpose computers, but are very limited in what they can do. The basic problem lies in the fact that the "memory" of an FSM is carried in the states it has. As such, finite state machines cannot be constructed for tasks for which we cannot set a fixed limit on the memory requirements. For example, the task of recognizing input sequences of the form A^nB^n (*i.e.*, a string of n A's followed by a string of n B's) *when we do not know what n is beforehand*, will be beyond the capabilities of FSMs. This is so, because we basically need a state to remember that we have seen k A's for $k = 0, 1, 2, ...,$ and all that is required to disprove any FSM purported to recognize A^nB^n is to present it with an input sequence using a value for n that is greater than the number of states it has.

We can solve this limitation by allowing our abstract machines to have access to memory. Figure 17 illustrates the general form of such *unrestricted tape automata*. The automaton has a *control unit* (a finite state device), a read only *input tape*, a write only *output tape*, a read/write *storage tape*, and a movable read and/or write *head* for each tape. Each tape is a storage medium divided into squares with one symbol per square. We imagine that a tape is oriented horizontally and that it extends indefinitely to the left and to the right. At any point in time, all but a finite sequence of adjacent tape squares are inscribed with the symbol '#' which signifies a blank square. Each step in the operation of a tape automaton consists of three actions: 1) the movement of a given head one square to the left or to the right along its tape; 2) the reading of the symbol found in the new square, or the writing of a specified symbol in the new square; and 3) the transition of the control unit to a new state.

Figure 17:

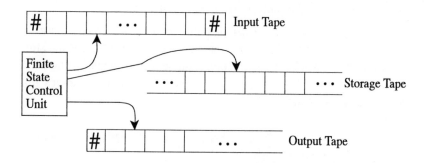

A *push-down automaton* (PDA) is a tape automaton with the following restrictions: 1) the machine is not allowed to move its input head to the left; and 2) it writes a symbol on the storage tape each time it moves the storage head to the right and it reads a symbol from the storage tape each time it moves the storage head to the left. Note that any symbol on the right of the storage head is irretrievable since it will be overwritten when the storage head again moves to the right. Although the storage tape gives PDA *unbounded memory*, access to memory is restricted by the fact that the information most recently written into the memory must be the first to be retrieved. This form of limited access storage is known as *push-down stack* because it implements a "last-in, first-out" retrieval rule.

We can readily construct a PDA to recognize input sequences of the form A^nB^n. Figure 18 gives the abbreviated transition diagram for such a PDA. In the diagram, an arc label of the form $a, b/cde$ indicates that if the input tape symbol read is a and the storage tape symbol read is b, then the symbols c, d, and e are written onto the storage tape. Note that the first action taken by the PDA, signified by /@ is to write a special symbol @ on the storage tape. The accepting state (also called the *suc-*

cess or *final* state) is s_1 and is denoted by a double circle.

Figure 18:

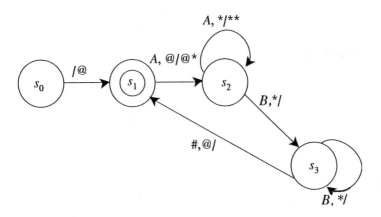

Although a PDA is more versatile than an FSM, it still has limitations. For example, it is not possible to construct a PDA to recognize input sequences of the form $A^n B^n C^n$; but, we can construct a *linear bounded automaton* (LBA) to do that.

Linear bounded automata form a much larger and more versatile class of computation models. An LBA is a tape automaton with the following restrictions: 1) the machine is not allowed to move its input head to the left; and 2) the length of the storage tape is some fixed multiple of the length of the input string. (The adjective "linear" is used in describing the machine because the amount of storage permitted is related linearly to the length of an input string.) Note that the read/write head is now allowed to move freely along the storage tape, reading (without erasing) as it goes and writing on any specified square.

Linear bounded automata represent an intermediate level of computational power between push-down automata and Turing machines.

7.3   TURING MACHINES

A Turing machine is a tape automaton with a single (storage) tape and the ability to read and write symbols on the storage tape with no restriction other than the head may only move one square at a time. We can describe the actions of any particular Turing machine by a set of quintuples of the form $<s, i, i', s', d>$ where s and i indicate the present state and the tape symbol being read, i' denotes the symbol printed, s' denotes the new state, and d signifies the direction in which the read/write head moves (R for right, L for left).

Example: What will result if the Turing machine described by,

$<0, 0, 1, 0, R>$

$<0, 1, 0, 0, R>$

$<0, \#, 1, 1, L>$

$<1, 0, 0, 1, R>$

$<1, 1, 0, 1, R>$,

is started with the input tape, #0110#?

By convention, the starting state is numbered 0 and the read/write head is assumed to be over the leftmost non-blank symbol on the tape. The different stages of the computation are as follows (the underlined symbol indicates the position of the head): ##1$\underline{1}$10##; ##10$\underline{1}$0##; ##100$\underline{0}$##; ##1001$\underline{\#}$#; #100$\underline{1}$1#; #1000$\underline{1}$#; #10000$\underline{\#}$. The machine halts because there are no quintuples describing the action to be taken in state 1 and reading a blank (#).

As we have seen, Turing machines are only one of many classes of abstract computing machines. However, it has been found that for each proposed class of automata, a Turing machine can be devised that will carry out the computations of any member of the class. In fact, there are several ways in which the basic Turing machine model can be modified without affecting its ultimate computing capabilities. Among the generalizations of the model that do not enhance its computing power are the use of multi-dimensional tapes, the use of more than one reading head, and even allowing for *non-deterministic* behavior (that is, allowing for a choice of next states). Among the restrictions that do not decrease the power of the basic model are the use of singly infinite tapes, the limitation to one non-blank tape symbol, or the limitation to two internal states. As such, the broad capabilities of the Turing machine suggest that it might represent an inherent limit on computational ability. Indeed, Turing proposed that *any* computation one might naturally regard as possible to carry out, can be performed by some Turing machine having a suitable set of instructions. This proposal, that Turing machines are a formal counterpart to the informal notion of algorithms, has become known as *Turing's thesis*.

We conclude this section by noting that we have considered Turing machines as special-purpose computers. To design a *general-purpose* Turing machine, we need to design a machine that can accept as input both the input string and a description of the computation that is to be done. This is, in fact, precisely what a general-purpose computer does; it accepts both the input data and the program which is a description of the computation to be done. We address the problem of constructing a similar general-purpose Turing machine in Chapter 9.

CHAPTER 8

FORMAL LANGUAGES

8.1 GRAMMARS AND LANGUAGES

An *alphabet* or *vocabulary V* is a finite, non-empty set of symbols. The set V^* is the set consisting of all finite strings of elements from the set V. There are many possible interpretations of the elements of V^*, depending on the nature of V. If we think of V as a set of "words," then V^* may be regarded as the set of all possible "sentences" formed from words in V. And thus, any subset of V^* may be considered a *language* over V.

We think of a language as a complete specification of three things. First, there must be a set V consisting of all "words" which are to be considered as being part of the language. Second, a subset of V^* must be designated as the set of "properly structured sentences" in the language. And third, there must be a way of determining the "meaning" of the properly structured sentences. The specification of the proper construction of sentences is called the *syntax* of the language, while the specification of the meaning of sentences is called the *semantics* of the language. In this chapter, we will not deal with semantics at all. We will examine the syntax of a class of languages called *phrase structure grammars*. Although these languages are not complex enough to encompass "real" languages such as English, they include most aspects of programming languages, and they are simple enough to be studied precisely since the syntax is determined by *substitution rules*.

A phrase structure grammar (*type 0 grammar*) G is a 4-tuple, G

= $<V, V_T, S, \rightarrow>$, where V is a vocabulary, V_T is a non-empty subset of V called the set of *terminals*, the *start symbol S* is a member of the set of *non-terminal symbols* (i.e., $V - V_T$), and $\rightarrow$ is a finite relation on V^* called the *productions* of G. The relation $\rightarrow$ is a *replacement* relation in the sense that if $w \rightarrow w'$, then we may replace w by w' whenever the string w occurs either alone or as a substring of some other string. Normally, $w \rightarrow w'$ is called a *production* of G, and we may refer to w and w', respectively, as the *left* and *right* sides of the production. We will assume that no production of G has the *empty string* (λ) as its left side.

Let us now define the relation $\Rightarrow$ on V^* as follows. We say $x \rightarrow y$, if $x = lwr$ and $y = lw'r$ and $w \rightarrow w'$, where l and r are completely arbitrary strings in V^*. In other words, $x \Rightarrow y$ means that y is *directly derivable* from x by using one of the productions of G to replace all or part of x. Finally, we let $\Rightarrow^\infty$ to be the *transitive closure* of $\Rightarrow$, and decree that a sentence w is *syntactically correct* if $w \varepsilon V_T^*$ and $S \Rightarrow^\infty w$.

The set of all syntactically correct (properly constructed) sentences that can be produced using a grammar G, is called the *language of G*, and is denoted by $L(G)$.

Example: A portion of formal grammar to generate identifiers in a programming language could be as follows:

identifier $\rightarrow$ letter

identifier $\rightarrow$ identifier letter

identifier $\rightarrow$ identifier digit

letter $\rightarrow a$

letter $\rightarrow b$

...

letter $\rightarrow z$

digit $\rightarrow 0$

digit $\rightarrow 1$

...

digit $\rightarrow 9$

Here, the set of non-terminals is {identifier, letter, digit}, V_T = {a, b, ..., z, 0, 1, ..., 9}, and the start symbol is "identifier."

A common shorthand known as *Backus-Naur form* (BNF) avoids listing all these productions as shown below:

<identifier> ::= <letter>|<identifier><letter>|<identifier><digit>

<letter> ::= $a \mid b \mid c \mid ... \mid z$

<digit> ::= $0 \mid 1 \mid 2 \mid ... \mid 9$.

Example: Let G be a phrase structure grammar with $V_T = \{a, b, c\}$, the set of non-terminals $\{S, w\}$, and the following productions:

1. $S \rightarrow aSb$
2. $Sb \rightarrow bw$
3. $abw \rightarrow c$

In order to determine $L(G)$, we start by noting that since we must start with S, we must use production 1 first and we may continue using it resulting in a string of the form a^nSb^n. Eventually, we must use the second production to eliminate S. This will result in strings of the form a^nwb^n. Now we must use production 3 to remove the non-terminal symbol w which finishes the substitution process. Therefore, $L(G)$ consists of strings of the form a^ncb^n, where $n \geq 0$.

8.2   CLASSES OF GRAMMARS

An erasing production is a production of the form $w \rightarrow \lambda$. To generate any language containing the empty string we have to be able to erase somewhere. In the following grammar types we limit erasing, if it occurs at all, to a single production of the form $S \rightarrow \lambda$ where S is the start symbol and if this production occurs we assume that S does not appear on the right side of any production. This is called the *erasing convention*.

A grammar is *context-sensitive* (*type 1*) if it obeys the erasing convention and for every production $w \rightarrow w'$ (except $S \rightarrow \lambda$), the word w' is at least as long as the word w. A grammar is *context-free* (*type 2*) if it obeys the erasing convention and for every production $w \rightarrow w'$, w is a single non-terminal symbol. A grammar is *regular* (*type 3*) if it obeys the erasing convention and for every production $w \rightarrow w'$ (except $S \rightarrow \lambda$), w is a single non-terminal symbol and w' is of the form tt' where t is a terminal symbol and t' is a non-terminal symbol. It follows from these definitions that each type of grammar is a special case of the type preceding it. This hierarchy of grammars, from the unrestricted type 0 to the most restricted type 3, is also called the *Chomsky hierarchy*.

A language is type 0 (context-sensitive, context-free, regular) if it can be generated by a type 0 (context-sensitive, context-free, regular) grammar.

We conclude this section by examining the *parsing* process. Parsing involves taking a sentence and verifying that it is syntactically correct in some grammar G by constructing a derivation that will produce it. (Because context-free grammars allow the replacement of only one symbol at a time, the derivation of a sentence can be represented as a *parse tree* (see Figure 19 for an example)). Given a sentence, *top-down parsing* attempts to construct a derivation for it by beginning with the start symbol, applying productions, and ending with the sentence. *Bottom-up parsing* begins with the sentence and tries to determine what productions were used to create it, applies the productions "backwards," and ending with the start symbol.

Figure 19:

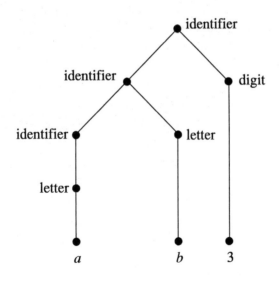

8.3 LANGUAGE RECOGNIZERS

We have now discussed four models of computation (finite state machines, push-down automata, linear bounded automata, Turing machines) and four types of languages (regular, context-free, context-sensitive, type 0). We have seen that as one moves through the hierarchy of computation models from FSMs to Turing machines, increasingly more complex computations can be performed. Similarly, as one moves from type 3 grammars to type 0 grammars, increasingly more complex sentences can be formed. Therefore, the following correspondence between machines and languages, where the ordered pair <machine, language> indicates that each machine of the given type accepts or recognizes some language of the given type, should not be too surprising:

<finite state machines, regular languages>

<*non-deterministic* push-down automata, context-free languages>

<*non-deterministic* linear bounded automata, context-sensitive languages>

<Turing machines, type 0 languages>

We conclude this section by answering a question we posed in the previous chapter. What properties must a set of input sequences have if it is to be *recognizable* by some finite state machine? The answer is that the most restricted class of languages coincides with the class of sets recognized by FSMs. In other words, finite state machines can recognize *regular expressions* which are defined as follows.

Let V be a vocabulary. The set of regular expressions on V is the set generated by these rules:

1. Each element in V is a regular expression, as are the expressions { } and λ, denoting the empty set and the empty string, respectively.

2. If P and Q are regular expressions, then so are the union $P \lor Q$ and the concatenation PQ.

3. If P is a regular expression, then so is its closure $(P)^*$.

4. The only regular expressions on V are those generated by finitely many applications of these rules.

Example: What is the set of strings on $V = \{a, b\}$ produced by the regular expression: $a(b \lor a)^*ba$?

The strings produced consist of "*a*" concatenated with a string produced by $(b + a)^*$, concatenated with "*ba*". The strings produced by $(b \lor a)^*$ are arbitrary mixes of "*b*" and "*a*". Therefore, the set of strings produced will be: *aba*; *abba*; *aaba*; *ababa*; *abbba*; *aabba*; *aaaba*; ....

Example: Consider the FSM, M, given by the state graph in Figure 20. The accepting state s_2 is denoted by a double circle. Write a regular expression describing the set recognized (accepted) by M.

The answer is $10(10)^*$.

Figure 20:

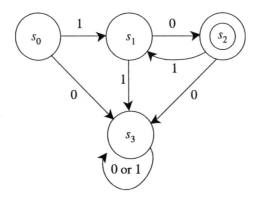

Example: Construct a FSM, M, to recognize the regular expression $a(b \lor a)^* ba$.

Figure 21:

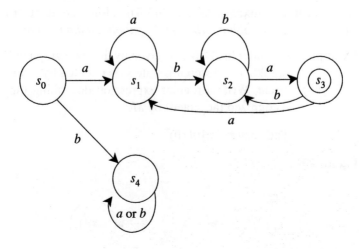

The answer is given in Figure 21.

CHAPTER 9

COMPUTABILITY

9.1 ALGORITHMS AND PROGRAMS

Computer scientists usually associate the word "algorithm" with computer programs. But, the informal idea of an algorithm is quite old, and, in fact, careful formulations of the idea were presented by mathematical logicians prior to the development of the first stored-program computer.

Computability theory is concerned with functions computed by algorithms, such as computer programs. These functions are called *computable functions*.

We consider an algorithm to work on input data to produce output data. Moreover, we expect that both the input to an algorithm, and the output from it, have finite descriptions. We also insist that an algorithm must lead to the same outcome every time it is started with the same input. As such, an algorithm establishes a *functional* relationship between its set of possible input data and its set of possible output data. Furthermore, we allow the function so defined to be a *partial* one; that is, in the same way that a computer program may not halt for a given input, the computation prescribed by an algorithm need not terminate for every choice of input data. Finally, we restrict our attention to algorithms that consist of finite numbers of instructions and therefore can be represented by finite descriptions.

Example: Let g be a total one-variable function such that $g(n + 1)$

$> g(n)$ for all $n \geq 0$. Now consider the function $f: N$ {0, 1}, where

$$f(x) = \begin{array}{l} 1 \quad \textit{if x is in range of g otherwise} \\ 0 \end{array}$$

Is f a computable function?

The following represents an *effective procedure* for determining the value of $f(x)$ for any natural number x. Evaluate $g(0)$, $g(1)$, $g(2)$, ... until finding the first argument k such that $g(k) \geq x$. (Since g is assumed to be strictly monotonic, such a value must eventually be reached.) Now, if $g(k) = x$, then x certainly belongs to the range of g, so $f(x) = 1$. If $g(k) > x$, then because g is monotonic, no larger argument value for g will yield x as a function value, and therefore x does not belong to the range of g, so $f(x) = 0$.

Example: Consider the function $f: N \to$ {0, 1}, where

$$f(x) = \begin{array}{l} 1 \quad \textit{if there exists a run of at least x} \\ \quad \textit{5's in the decimal expansion of } \pi \\ 0 \quad \textit{otherwise} \end{array}$$

Is f a computable function?

It is relatively easy to see that f is either the constant function 1 (that is, for all x, there exists a run of *at least* x consecutive 5's in the decimal expansion of π), or it is the step function (for some k, $f(x) = 1$ if $x < k$ and $f(x) = 0$ for $x \geq k$, reflecting the situation that there exists a maximum run of consecutive 5's in the decimal expansion of π). In either case, f is equivalent to a computable function (the constant or the step functions), and therefore is considered to be computable.

What is an effective procedure for computing $f(x)$? The surprising answer is that we do not know! The problem is essentially that we do not know *when* to conclude that $f(x) = 0$ for some x.

Now, let us consider the question of actual specification of algorithms. Several independent formalisms have been proposed including the Turing machine description of algorithms, the Kleene characterization of computable functions, Post production systems, and Church's lambda-calculus. Each formalism yields a corresponding class of computable functions. And, by a hypothesis known as *Church's thesis*, each formalism is considered to be *complete*, that is, every conceivable algorithm, independent of its form of expression, must be equivalent to some algorithm specified in that formalism.

9.2   THE UNIVERSAL TURING MACHINE

Can we construct a single Turing machine that will simulate the actions of any Turing machine on any input? The answer is yes and such a machine is called a *universal Turing machine*. In order to carry out its computations, a universal Turing machine must be provided with two things: a description of the machine it is to simulate, and a description of that machine's initial tape pattern. Since every machine can be represented by a set of ordered quintuples, and since every quintuple can in turn be represented by a series of blocks of 1's, machines can be described by strings of 0's and 1's. A similar coding scheme can also be used for the description of arbitrary tape patterns. Once these representational schemes have been decided upon, the design of the control unit of the universal machine itself is straightforward. The fact that it is possible to design a universal Turing machine also follows from Church's thesis, because there exists an effective procedure that runs any Turing machine on any input. Specifically, the process of "look at tape symbol and present state, examine the list of quintuples until you find the right match and do what it indicates; if you do not find the right quintuple, halt" is such an effective procedure.

9.3   UNSOLVABILITY

By Church's thesis, Turing machines can do a great deal indeed. However, of more interest than the question of what Turing machines can do, is the question of what it is that they *cannot* do. This is so because unbounded memory has given Turing machines abilities that exceed those of an actual computer, and if we find something that no Turing machine can do, then a real computer cannot do it either. And, by Church's thesis, no algorithm exists to do it as well.

The *halting problem* for Turing machines is an example of a class of problems known as *decision problems*. A decision problem is characterized by two things: a set of objects S, and an n-ary relation P on the set S. The decision problem itself is that of determining, for any arbitrary combination of n objects from S, whether or not the relation P holds for that combination. A decision problem is said to be *effectively decidable* if and only if there exists an effective procedure that correctly determines, for every possible combination of n objects from S, whether or not the requisite relation (property) P holds for that combination. A decision problem is said to be *undecidable* (or *unsolvable*) if and only if it can be proven that no such effective procedure exists.

Example: The following are instances of decision problems.

a) The problem of determining, for any pair of natural numbers, whether those numbers are relatively prime. Here, S is the set of natural numbers N and P is the binary relation that holds between two numbers if and only if they are relatively prime.

This problem is effectively decidable since the familiar Euclidean algorithm provides an effective way for making the necessary decision.

b) Hilbert's Tenth Problem is the problem of determining, for any polynomial equation $P(x_1, x_2, ..., x_n) = 0$ with integral coefficients, whether integral solutions exist. Here, S is the set of all polynomial equations with integral coefficients, and P is the unary relation that holds for a polynomial equation if and only if it has integral solutions.

Although formulated in 1900, it took until 1970 for the problem to be proven as unsolvable.

The halting problem for Turing machines asks: does an algorithm exist to decide, given any Turing machine T and string w, whether T begun on a tape containing w will eventually halt? We can prove the unsolvability of the halting problem as follows:

Step 1. Assume that the halting problem is solvable. Let X be the Turing machine that decides the halting problem. Specifically, let us assume that given the description of machine T, s_T, and string w, X will halt with an output of 1 if and only if T halts when started with w, otherwise X will halt with an output of zero.

Step 2. Let us now construct a new machine, Y, as follows. We add to X's quintuples enough quintuples to modify its behavior as follows. When ordinarily X should halt with an output of 1, the new quintuples will take Y to a new state which will cause it to move right continuously, never halting. So, Y begun on $<s_T, w>$ will halt with an

76

output of zero if and only if X halts with an output of zero. However, Y begun on $<s_T, w>$ will not halt if and only if X halts with an output of 1.

Step 3. We now construct a new machine, Z, as follows. When started with an input string w', Z first copies w' and then turns control over to Y. Effectively, Z simulates the behavior of Y on $<w',w'>$.

Step 4. Now, consider what happens if we start Z with an input string that is its own description, s_z. There are two possibilities:

Case i. Z halts when started with s_z. By the way Z has been constructed, Z halts if Y halts on $<s_z, s_z>$. By the way Y has been constructed, Y halts if X halts with an output of zero when started on $<s_z, s_z>$. But, by definition of X, X halts with an output of zero when started on $<s_z, s_z>$ if and only if the Turing machine defined by s_z, i.e., Z, does *not* halt when started with s_z. Thus, Z halts when started with s_z if Z does not halt when started with s_z – a contradiction.

Case ii. Z does not halt when started with s_z. By the way Z has been constructed, Z does not halt if Y does not halt on $<s_z, s_z>$. By the way Y has been constructed, Y does not halt if X halts with an output of 1 when started on $<s_z, s_z>$. But, by definition of X, X halts with an output of 1 when started on $<s_z, s_z>$ if and only if the Turing machine defined by s_z, i.e., Z, *does* halt when started with s_z. Thus, Z does not halt when started with s_z if Z does halt when started with s_z – a contradiction.

Together, these two cases provide a contradiction which refutes our assumption that a machine X exists. Thus, there is no effective procedure for deciding the halting problem.

We conclude this section by examining the *blank tape halting problem*. Specifically, does an algorithm exist to decide, given any Turing machine T, whether T started on a blank tape will eventually halt? To answer this question, we resort to the powerful notion of *reducibility*. Suppose that A and B are two decision problems. If by assuming that B is solvable, we can deduce that A is also solvable, then we say that *A is reducible to B*. Now, if we already know that A is unsolvable, then so must B.

To prove that the blank tape halting problem is unsolvable, we will reduce the general halting problem to it. Suppose that we have a machine X that solves the blank tape halting problem (i.e., when started with a description of Turing machine T, it will halt with an output of 1 if T halts on a blank tape, otherwise it will halt with an output of zero). Let T be an arbitrary Turing machine and let w be an arbitrary string, and consider the construction of a Turing machine T', which, when started on blank tape, writes w on its tape and then simulates T. Obviously, T' halts when begun on blank tape if and only if T halts when started on w. Thus, we may construct a Turing machine X' that solves the halting problem as follows: given a description of T and w, X' constructs a description of the corresponding machine T' on its tape, erases the rest of its tape, and passes control to X. Clearly, X' halts with 1 or 0 according as T does or does not halt for input w. Therefore, the general halting problem is reducible to the blank tape halting problem.

CHAPTER 10

SOME APPLICATION AREAS

10.1   LOGIC CIRCUIT DESIGN

The Boolean algebra operations of addition, multiplication, and complementation can be associated with electronic devices known as the *OR gate*, the *AND gate*, and the *NOT gate* (or the *inverter*). Figure 22 shows the standard symbols for these circuits. In theory, an input or output wire is assigned a value of 1 or 0 (corresponding to True and False, respectively), depending on its voltage. In actuality, because of signal fluctuations, a range of voltage values is associated with 1 and another range with 0. Also, it helps to assume that the time it takes for the device to produce its output is negligible. A digital circuit consisting of such devices is called a *combinatorial logic network*.

Figure 22:

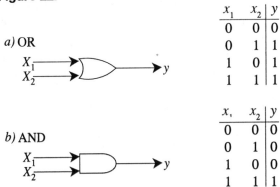

x_1	x_2	y
0	0	0
0	1	1
1	0	1
1	1	1

a) OR

x_1	x_2	y
0	0	0
0	1	0
1	0	0
1	1	1

b) AND

c) NOT

x_1	y
0	1
1	0

Consider the problem of designing the logic circuitry for a typical seven segment LED (light emitting diode) display (see Figure 23a) to be used for displaying the digits 0 through 9. The inputs to the required circuit are the four bits representing the digit to be displayed, and the outputs are the status (1 and 0 to denote respectively on and off) of each of the seven LED segments. The truth table below represents the desired behavior.

x_3	x_2	x_1	x_0	digit	y_1	y_2	y_3	y_4	y_5	y_6	y_7
0	0	0	0	0	1	0	1	1	1	1	1
0	0	0	1	1	0	0	0	0	1	0	1
0	0	1	0	2	1	1	1	0	1	1	0
0	0	1	1	3	1	1	1	0	1	0	1
0	1	0	0	4	0	1	0	1	1	0	1
0	1	0	1	5	1	1	1	1	0	0	1
0	1	1	0	6	0	1	1	1	0	1	1
0	1	1	1	7	1	0	0	0	1	0	1
1	0	0	0	8	1	1	1	1	1	1	1
1	0	0	1	9	1	1	0	1	1	0	1

The Boolean expression for output y_1 is derived by noticing from the truth table that y_1 will be 1 if either of the following (product) terms is 1: $x_3'x_2'x_1'x_0'$, $x_3'x_2'x_1 x_0'$, $x_3'x_2'x_1 x_0$, $x_3'x_2 x_1$, x_0, $x_3'x_2 x_1 x_0$, $x_3 x_2'x_1'x_0'$, $x_3 x_2'x_1'x_0$. Therefore, y_1 can be represented by the sum (OR) of the preceding items. Further simplification results in: $y_1 = x_3 + x_2'x_1 + x_2'x_0' + x_2 x_0$. Figure 23 gives the circuit diagram for realizing y_1.

Figure 23:

a) Seven Segment LED Display

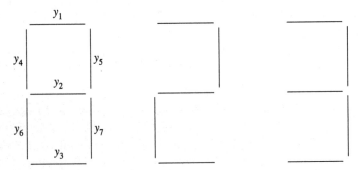

b) Combinational Circuit to Control y_1

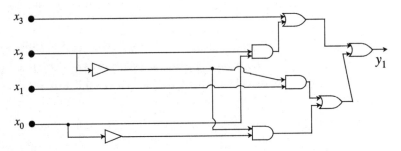

10.2   RELATIONAL DATABASES AND RELATIONAL ALGEBRA

In relational database terminology, the term relation is used to refer to both the *intention* (data type) of a relation as well as its *extension* (occurrence or instance). Each relation instance is a subset of a Cartesian product, $D_1 \underline{x} D_2 \underline{x} \ldots \underline{x} D_n$, where the D_i are sets of values called the domains of the relation. A *domain* is a set of values specified by a name, and the Cartesian product of a set of domains, $D_1, D_2, \ldots, D_n$, is the set of all *tuples* $<V_1, V_2, \ldots, V_n>$, where V_I is a value belonging to domain D_1,

V_2 is a value belonging to domain D_2, and V_n is a value belonging to domain D_n. Figure 24 depicts both the intention and an extension of a relational database about suppliers, parts, parts currently being supplied, and parts expected to be supplied. As it can be seen, it is helpful to think of a relation as a table of data. Each column is identified by an attribute name in addition to a domain specification. Each row corresponds to an n-tuple (the n refers to the *degree* of the relation, i.e., the number of domains or columns over which it is defined). Because each relation is a set, the ordering of rows is immaterial and all rows are distinct.

A relational algebra provides a set of operators that can be used to construct any desired relation from those in the database and any intermediate relations that may be required. The union of two relations R and S, denoted (R) **UNION** (S), is the set of tuples belonging to either R or S or both. The intersection of the two relations, (R) **INTERSECTION** (S), consists of those tuples belonging to both R and S. Finally, the set of tuples belonging to R and not to S defines the difference between R and S, (R) **DIFFERENCE** (S). (See Figure 25 for examples of these and other relational operators.)

The projection of relation R on the attribute list A is defined as the set:

PROJECTION$_{[A]}$ $(R) = \{\ r.A \mid r \in R\}$, where r denotes an element (a row) of R and $r.A$ represents a tuple (row) constructed from r by keeping only those column values specified by attribute list A.

The selection operator is used to select those tuples in a relation that satisfy a selection condition. The selection predicate consists of simple conditions of the form $A\ ec$ combined with Boolean connectives, where A is an attribute of the relation, e is an arithmetic comparison operator ($<$, $>$, $\leq$, $\geq$, $=$, and $\neq$), and c is a constant value from the domain of A. In addition, it is possible to specify a condition involving two compatible attributes of the same relation, e.g., $A\ e\ B$. More formally, the selection on relation R using the selection predicate, is defined as:

SELECTION$_{[\Phi]}$ $(R) = \{\ r \mid (r \in R)\ \textbf{AND}\ \Phi,(r)\ \}$.

The restriction (semijoin) operator allows us to obtain a subset of a relation based on a qualification involving another relation. Formally,

(R) **RESTRICTION**$_{[A\ IN\ B]}$ $(S) = \{\ r \mid (r \in R)\ \textbf{AND}\ (r.A \in \textbf{PROJECTION}_{[B]}$ $(S))\ \}$, or

(R) **RESTRICTION**$_{[A \text{ NOT IN } B]}$ $(S) = \{\ r\ |\ (r\ \varepsilon\ R)$ **AND NOT** $(r.A\ \varepsilon$
PROJECTION$_{[B]}$ $(S))\ \}$.

Consider now the following data selection operation. Group the rows of relation R based on their values for attribute A. Select each group whose associated *set of values* for attribute B contains *every* distinct value in column Z of relation S. In relational algebra, this operation is performed by the *generalized division* operator:

(R) **GD**$_{[(A)\ B \text{ THETA } Z]}$ $(S) =$

$\{\ r\ |\ (r\ \varepsilon\ R)$ **AND** $(($**PROJECTION**$_{[B]}$ $($**SELECTION**$_{[A\ =\ r.A]}$ $(R)))$
THETA $($**PROJECTION**$_{[Z]}$ $(S)))\}$, where **THETA** represents one of the following set comparison operators: IS EQUAL TO, IS NOT EQUAL TO, CONTAINS, DOES NOT CONTAIN, IS IN, and IS NOT IN. For example, (FUTURE) **GD**$_{[(S\#)\ P\# \text{ CONTAINS } P\#]}$ (NOW) produces a report listing $S\#$, and $P\#$ for all suppliers who can supply in the future all parts currently being supplied.

The *grouped generalized division* operator provides the capacity to compare sets of values associated with *matching* groups of tuples in two relations. It is defined formally by:

(R) **GGD**$_{[(A)\ B \text{ THETA } Z\ (Y)]}$ $(S) =$

$\{\ r\ |\ (r\ \varepsilon\ R)$ **AND** $(\ ($**PROJECTION**$_{[B]}$ $($**SELECTION**$_{[A\ =\ r.A]}$ $(R)))$

THETA

$($**PROJECTION**$_{[Z]}$ $($**SELECTION**$_{[Y\ =\ r.A]}$ $(S))))\ \}$,

where attribute lists A and Y are known as the grouping attributes and attribute lists B and Z are referred to as the division attributes. For example,

(FUTURE) **GGD**$_{[(S\#)\ P\# \text{ EQUALS } P\# (S\#)]}$ (NOW) produces a report listing $S\#$, and $P\#$ for those suppliers who are expected to supply exactly the same set of parts they currently supply.

The join of two relations R and S is constructed by taking each tuple from relation R, determining if it satisfies the join condition with each tuple of S and concatenating them if it does. The join condition is expressed as $A\ \Theta\ B$, where A and B are, respectively, union-compatible attributes of R and S, and Θ is an arithmetic comparison operator. More formally, the Θ-join operator is defined as:

(R) **JOIN**$_{[A\ \Theta\ B]}$ $(S) = \{\ r^{\wedge}s\ |\ (r\ \varepsilon\ R)$ **AND** $(s\ \varepsilon\ S)$ **AND** $(r[A]\ \varepsilon\ s[B])\ \}$, where $r^\wedge s$ denotes the tuple constructed by concatenating r with s.

We conclude this section by noting that the relational operators defined above are not independent; that is, some can be expressed in terms of the others. For instance, the Θ-join operation can be expressed by the Cartesian product and the selection operations as the following identity indicates:

$$(R) \text{ } \underline{\textbf{JOIN}}_{[A \Theta B]} (S) = \underline{\textbf{SELECTION}}_{[A \Theta B]} (R \text{ } \underline{x} \text{ } S)$$

Figure 24:

a) Relational Schema (relation intentions)

```
SUPPLIER  (S#, SNAME, CITY)
PART      (P#, PNAME, COLOR)
NOW       (S#, P#)
FUTURE    (S#, P#)
```

b) Sample Database Instance   (relation extensions)

S#	SNAME	CITY
S1	SMITH	LONDON
S2	JONES	PARIS
S3	BLAKE	PARIS
S4	CLARK	LONDON
S5	ADAMS	ATHENS
S6	ROBERTS	PARIS

Supplier

P#	PNAME	COLOR
Pl	NUT	RED
P2	BOLT	GREEN
P3	SCREW	BLUE
P4	SCREW	RED
P5	CAM	BLUE
P6	COG	RED
P7	BOLT	RED

Part

84

S#	P#	
S1	P1	
S2	P3	
S2	P5	
S3	P3	
S3	P4	
S4	P6	**Now**
S5	P1	
S5	P2	
S5	P3	
S5	P4	
S5	P5	
S5	P6	

S#	P#	
S2	P3	
S3	P4	
S4	P6	
S4	P7	**Future**
S5	P5	
S5	P6	
S6	P1	
S6	P7	

Figure 25:

(NOW) INTERSECTION (FUTURE)

S#	P#
S2	P3
S3	P4
S4	P6
S5	P5
S5	P6

(NOW) DIFFERENCE (FUTURE)

S#	P#
S1	P1
S2	P5
S3	P3

S5   P1
S5   P2
S5   P3
S5   P4

(SUPPLIER) RESTRICTION$_{[s\# \text{ NOT IN } s\#]}$ (FUTURE)

S#	SNAME	CITY
S1	SMITH	LONDON

(FUTURE) JOIN$_{[P\#=P\#]}$ (PART)

S#	P#	P#	PNAME	COLOR
S2	P3	P3	SCREW	BLUE
S3	P4	P4	SCREW	RED
S4	P6	P6	COG	RED
S4	P7	P7	BOLT	RED
S5	P6	P6	COG	RED
S5	P5	P5	CAM	BLUE
S6	P1	P1	NUT	RED
S6	P7	P7	BOLT	RED

SELECTION$_{[\text{COLOR} = \text{'BLUE'}]}$ (PART)

P#	PNAME	COLOR
P3	SCREW	BLUE
P5	CAM	BLUE

(NOW) GD$_{[(s\#) \text{ P\# CONTAINS P\#}]}$ (SELECTION$_{[\text{COLOR} = \text{'BLUE'}]}$ (PART))

S#	P#
S2	P3
S2	P5
S5	P1

S5	P2
S5	P3
S5	P4
S5	P5
S5	P6

(NOW) GDG$_{[(S\#)\ P\#\ CONTAINS\ P\#\ (S\#)]}$ (FUTURE)

S#	P#
S2	P3
S2	P5
S5	P1
S5	P2
S5	P3
S5	P4
S5	P5
S5	P6

(NOW) GDG$_{[(P\#)\ S\#\ CONTAINS\ S\#\ (P\#)]}$ (FUTURE)

S#	P#
S2	P3
S3	P3
S3	P4
S5	P3
S5	P4

REA's **Problem Solvers**

The "PROBLEM SOLVERS" are comprehensive supplemental text-books designed to save time in finding solutions to problems. Each "PROBLEM SOLVER" is the first of its kind ever produced in its field. It is the product of a massive effort to illustrate almost any imaginable problem in exceptional depth, detail, and clarity. Each problem is worked out in detail with step-by-step solution, and the problems are arranged in order of complexity from elementary to advanced. Each book is fully indexed for locating problems rapidly.

ADVANCED CALCULUS	HEAT TRANSFER
ALGEBRA & TRIGONOMETRY	LINEAR ALGEBRA
AUTOMATIC CONTROL	MACHINE DESIGN
SYSTEMS/ROBOTICS	MATHEMATICS for ENGINEERS
BIOLOGY	MECHANICS
BUSINESS, MANAGEMENT, & FINANCE	NUMERICAL ANALYSIS
CALCULUS	OPERATIONS RESEARCH
CHEMISTRY	OPTICS
COMPLEX VARIABLES	ORGANIC CHEMISTRY
COMPUTER SCIENCE	PHYSICAL CHEMISTRY
DIFFERENTIAL EQUATIONS	PHYSICS
ECONOMICS	PRE-CALCULUS
ELECTRICAL MACHINES	PSYCHOLOGY
ELECTRIC CIRCUITS	STATISTICS
ELECTROMAGNETICS	STRENGTH OF MATERIALS &
ELECTRONIC COMMUNICATIONS	MECHANICS OF SOLIDS
ELECTRONICS	TECHNICAL DESIGN GRAPHICS
FINITE & DISCRETE MATH	THERMODYNAMICS
FLUID MECHANICS/DYNAMICS	TOPOLOGY
GENETICS	TRANSPORT PHENOMENA
GEOMETRY	VECTOR ANALYSIS

*If you would like more information about any of these books,
complete the coupon below and return it to us or go to your local bookstore.*

RESEARCH & EDUCATION ASSOCIATION
61 Ethel Road W. • Piscataway, New Jersey 08854
Phone: (908) 819-8880

Please send me more information about your Problem Solver Books

Name _____

Address _____

City _____ State _____ Zip _____

REA's **Test Preps**
The Best in Test Preparations

dvanced Placement Exams
iology
alculus AB & Calculus BC
hemistry
nglish Literature & Composition
uropean History
nited States History

ollege Board Achievement Tests
merican History
iology
hemistry
nglish Composition
rench
ierman
panish
iterature
Iathematics Level I & II

Graduate Record Exams
Biology
Chemistry
Computer Science
Economics
Engineering
General
Literature in English
Mathematics
Physics
Psychology

FE - Fundamentals of Engineering Exam
GMAT - Graduate Management Admission Test
MCAT - Medical College Admission Test
NTE - National Teachers Exam
SAT - Scholastic Aptitude Test
LSAT - Law School Admission Test
TOEFL - Test of English as a Foreign Language
